Environmental Health Criteria 151

SELECTED SYNTHETIC ORGANIC FIBRES

First draft prepared by Dr M.E. Meek,
Environmental Health Directorate, Health
and Welfare, Ottawa, Canada

Published under the joint sponsorship of
the United Nations Environment Programme,
the International Labour Organisation,
and the World Health Organization

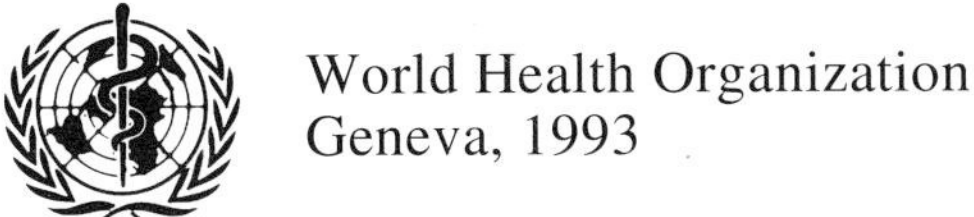

World Health Organization
Geneva, 1993

The **International Programme on Chemical Safety (IPCS)** is a joint venture of the United Nations Environment Programme, the International Labour Organisation, and the World Health Organization. The main objective of the IPCS is to carry out and disseminate evaluations of the effects of chemicals on human health and the quality of the environment. Supporting activities include the development of epidemiological, experimental laboratory, and risk-assessment methods that could produce internationally comparable results, and the development of manpower in the field of toxicology. Other activities carried out by the IPCS include the development of know-how for coping with chemical accidents, coordination of laboratory testing and epidemiological studies, and promotion of research on the mechanisms of the biological action of chemicals.

WHO Library Cataloguing in Publication Data

Selected synthetic organic fibres.

(Environmental health criteria ; 151)

1.Carbon - adverse effects   2.Nylons - adverse effects
3.Polyenes - adverse effects   4.Environmental exposure
I.Series

ISBN 92 4 157151 9   (NLM Classification: QV 627)
ISSN 0250-863X

PRINTED IN FINLAND
93/9804 – VAMMALA – 5500

CONTENTS

**ENVIRONMENTAL HEALTH CRITERIA FOR
SELECTED SYNTHETIC ORGANIC FIBRES**

WHO TASK GROUP ON ENVIRONMENTAL HEALTH CRITERIA FOR SELECTED SYNTHETIC ORGANIC FIBRES

Members

Dr D.M. Bernstein, Geneva, Switzerland

Dr J.M. Dement, Office of Occupational Health and Technical Services, National Institute of Environmental Health Sciences, Research Triangle Park, North Carolina, USA

Dr P.T.C. Harrison, Department of the Environment, London, United Kingdom

Professor T. Higashi, Department of Work Systems & Health, Institute of Industrial Ecological Sciences, University of Occupational & Environmental Health, Kitakyishu-shi, Japan

Dr I. Mangelsdorf, Institute for Toxicology, GSF München, München-Neuherberg, Germany

Dr E. McConnell, Raleigh, North Carolina, USA

Mrs M. Meldrum, Health and Safety Executive, Bootle, Merseyside, United Kingdom

Professor M. Neuberger, Institute of Environmental Hygiene, University of Vienna, Kinderspitalgasse, Vienna, Austria (*Chairman*)

Dr R.P. Nolan, Environmental Sciences Laboratory, Brooklyn College, Brooklyn, New York, USA

Dr V. Vu, Health Effects Branch, Health and Environmental Review Division, Office of Pollution Prevention and Toxics, US Environmental Protection Agency, Washington, DC, USA

Mr R. Zumwalde, Division of Standards Development and Technology Transfer, National Institute for Occupational Safety and Health, Robert A. Taft Laboratories, Cincinnati, Ohio, USA

Observers

Dr T. Hesterberg, Health Safety and Environment Department,
Mountain Technical Center, Schuller International Inc.,
Littleton, Colorado, USA

Dr E.A. Merriman, DuPont Fibers, Wilmington, Delaware, USA

Dr J.W. Rothuizen, Rothuizen Consulting, Genolier,
Switzerland

Dr D.B. Warheit, E.I. Du Pont de Nemours and Co., Haskell
Laboratory for Toxicology and Industrial Medicine,
Newark, Delaware, USA

Secretariat

Dr M.E. Meek, Environmental Health Directorate, Health
Protection Branch, Health and Welfare, Tunney's Pasture,
Ottawa, Ontario, Canada (*Rapporteur*)

Professor F. Valić, IPCS Consultant, World Health Organization,
Geneva, Switzerland; *also* Vice-Rector, University of
Zagreb, Zagreb, Croatia (*Responsible Officer* and *Secretary*)

NOTE TO READERS OF THE CRITERIA MONOGRAPHS

Every effort has been made to present information in the criteria monographs as accurately as possible without unduly delaying their publication. In the interest of all users of the Environmental Health Criteria monographs, readers are kindly requested to communicate any errors that may have occurred to the Director of the International Programme on Chemical Safety, World Health Organization, Geneva, Switzerland, in order that they may be included in corrigenda.

* * *

A detailed data profile and a legal file can be obtained from the International Register of Potentially Toxic Chemicals, Case postale 356, 1219 Châtelaine, Geneva, Switzerland (Telephone No. 9799111).

* * *

This publication was made possible by grant number 5 U01 ES02617-14 from the National Institute of Environmental Health Sciences, National Institutes of Health, USA.

ENVIRONMENTAL HEALTH CRITERIA FOR SELECTED SYNTHETIC ORGANIC FIBRES

A Task Group on Environmental Health Criteria for Selected Synthetic Organic Fibres met at the British Industrial and Biological Research Association (BIBRA), Carshalton, Surrey, United Kingdom, from 28 September to 2 October 1992. Dr. S.E. Jaggers opened the meeting on behalf of the host institute and greeted the participants on behalf of the Department of Health. Professor F. Valić welcomed the participants on behalf of the heads of the three cooperating organizations of the IPCS (UNEP/ILO/WHO). The Task Group reviewed and revised the draft monograph, made an evaluation of the direct and indirect risks for human health from exposure to the synthetic organic fibres reviewed, and made recommendations for health protection and further research.

The first draft was prepared by Dr M.E. Meek, Environmental Health Directorate, Health and Welfare, Ottawa, Canada. Professor F. Valić was responsible for the overall scientific content and for the organization of the meeting, and Dr P.G. Jenkins, IPCS, for the technical editing of the monograph.

ABBREVIATIONS

CKSCC	cystic keratinizing squamous cell carcinoma
FEV	forced expiratory volume
GC	gas chromatography
MMAD	mass median aerodynamic diameter
MMMF	man-made mineral fibre
MS	mass spectrophotometry
PAN	polyacrylonitrile
PCOM	phase contrast optical microscopy
SEM	scanning electron microscopy
SPF	specific pathogen free
TGA	thermogravimetric analysis
TWA	time-weighted average

INTRODUCTION

This document is a review of occupational and environmental exposure to selected synthetic organic fibres. The particular fibres covered in this document have been selected because they are the only ones for which some toxicological data are available, and they are probably of most importance in terms of production volumes and potential for human exposure. Considerations relating to possible future uses and applications of synthetic organic fibres have not been discussed, nor have the possible consequences of exposure to secondary products, combustion products, etc. The Task Group noted the complexity of the working environment in the manufacture and use of synthetic organic fibres, which may involve exposures to a range of chemical substances and non-fibrous dusts.

1. SUMMARY

1.1 Identity, physical and chemical properties

Carbon/graphite fibres are filamentary forms of carbon produced by high-temperature processing of one of three precursor materials: rayon (regenerated cellulose), pitch (coal tar or petroleum residue) or polyacrylonitrile (PAN). Nominal diameters of carbon fibres range between 5 and 15 μm. Carbon fibres are flexible and electrically and thermally conductive, and in high performance varieties have high Young's modulus (coefficient of elasticity measuring the softness or stiffness of the material) and tensile strength. They are corrosion resistant, lightweight, refractile and chemically inert (except to oxidation), and have a high degree of stability to traction forces, low thermal expansion and density, and high abrasion and wear resistance.

Aramid fibres are formed by the reaction of aromatic diamines and aromatic diacid chlorides. They are produced as continuous filaments, staple and pulp. There are two main types of aramid fibres, para- and meta-aramid, both with a nominal diameter of 12-15 μm. Para-aramid fibres can have fine-curled, tangled fibrils within the respirable size range (< 1 μm in diameter) attached to the surface of the core fibre. These fibrils may be detached and become airborne upon abrasion during manufacture or use. Generally, aramid fibres exhibit medium to very high tensile strength, medium to low elongation, and moderate to very high Young's modulus. They are resistant to heat, chemicals and abrasion.

Polyolefin fibres are long-chain polymers composed of at least 85% by weight of ethylene, propylene or other olefin units; polyethylene and polypropylene are used commercially. Except for some types such as microfibre, the nominal diameters of most classes of polyolefin fibres are sufficiently large that few are within the respirable size range.

Polyolefins are extremely hydrophobic and unreactive. Their tensile strengths are considerably less than those of carbon or aramid fibres and they are relatively flammable, melting at temperatures between 100 and 200 °C.

Methods developed for counting mineral fibres have been used for industrial hygiene monitoring of synthetic organic fibres. However, these methods have not been validated for this purpose.

Factors such as electrostatic properties, solubility in mounting media and refractive index may cause difficulties when using such methods.

1.2 Sources of human and environmental exposure

The estimated worldwide production of carbon and graphite fibres was in excess of 4000 tonnes in 1984. For aramid it was more than 30 000 tonnes in 1989, and for polyolefin fibres more than 182 000 tonnes (USA only). Carbon and aramid fibres are used principally in advanced composite materials in aerospace, military and other industries to improve strength, stiffness, durability, electrical conductivity or heat resistance. Polyolefin fibres are typically used in textile applications.

Exposures to synthetic organic fibres have been documented in the occupational environment. Synthetic organic fibres can be released into the environment during production, processing or combustion of composites and during disposal. Very few data are available concerning actual releases of these materials into the environment.

Available data on the transport, distribution and transformation of organic fibres in the environment are restricted to identification of products of municipal incineration of refuse from carbon fibre-containing composites and pyrolysis decomposition products of carbon fibre and aramids. During simulation of municipal incineration, both the diameters and lengths of carbon fibres were reduced. Principal pyrolysis decomposition products of carbon and aramid fibres include aromatic hydrocarbons, carbon dioxide, carbon monoxide and cyanides.

1.3 Environmental levels and human exposure

Synthetic organic fibre dusts are released in the workplace during operations such as fibre forming, winding, chopping, weaving, cutting, machining and composite formation and handling.

In the case of carbon/graphite fibres, respirable fibre concentrations are generally less than 0.1 fibres/ml but concentrations of up to 0.3 fibres/ml have been measured close to chopping or winding operations. Fibres may also be released during machining (drilling, sawing, etc.) of carbon fibre composites, although most of the respirable material thus produced is non-fibrous.

Average airborne concentrations of para-aramid fibrils in the workplace are reported to be less than 0.1 fibrils/ml in filament operations and less than 0.2 fibrils/ml in floc cutting and pulp operations. During staple yarn processing, average airborne fibril concentrations are typically below 0.5 fibrils/ml, but levels as high as approximately 2.0 fibrils/ml have been reported. Other end-use workplace exposures are typically below 0.1 fibrils/ml on average with peak exposures of 0.3 fibrils/ml. Special potential for exposure was demonstrated for waterjet cutting of composites, levels being as high as 2.91 fibrils/ml. Particles of mean aerodynamic diameter of 0.21 μm have been generated during laser cutting of epoxy plastics reinforced with aramid fibres, but the fibre content of the dust was not reported. Certain volatile organic compounds (including benzene, toluene, benzonitrile and styrene) and other gases (hydrogen cyanide, carbon monoxide and nitrogen dioxide) are also produced during such operations.

Limited air monitoring data from a facility producing polypropylene fibres indicate maximum airborne levels for fibres longer than 5 μm of 0.5 fibres/ml, with most values being less than 0.1 fibres/ml. Scanning electron microscopy showed that airborne fibre sizes range from 0.25 to 3.5 μm in diameter and 1.7 to 69 μm in length. In a single ambient sample collected near a carbon fibre weaving plant, a concentration of 0.0003 fibres/ml was detected. The average dimensions of the fibres were 706 μm by 3.9 μm. The release of carbon fibre at the crash site of two military aircraft, following combustion of carbon fibre composite used in construction, has also been reported. No other relevant information on concentrations in the environment was identified.

1.4 Deposition, clearance, retention, durability and translocation

Few data on specific synthetic organic fibres have been identified. The data on para-aramid fibres (Kevlar) indicate that, when inhaled, these fibres are deposited at alveolar duct bifurcations. There is also evidence of translocation to the tracheobronchial lymph nodes.

1.5 Effects on experimental animals and *in vitro* test systems

For the synthetic organic fibre types reviewed here, there is a dearth of good quality data from relevant experimental studies.

There are no adequate studies in which the fibrogenic or carcinogenic potential of carbon/graphite fibres has been examined. Effects following short-term inhalation exposure (days) of rats to respirable-size pitch-based fibres included inflammatory responses, increased parenchymal cell turnover and minimal type II alveolar cell hyperplasia. Available data from an intratracheal instillation and an intraperitoneal injection study are considered inadequate for assessment owing to the lack of characterization of the test materials and lack of adequate documentation of protocol and results. A mouse skin painting study on four carbon fibre types suspended in benzene was inadequate for the evaluation of carcinogenic activity.

In the case of paraaramid fibres, the majority of data is derived from experiments on Kevlar. Short-term (2 week) inhalation studies of Kevlar dust have resulted in a pulmonary macrophage response which decreased in severity after exposure ceased. Short-term studies of ultrafine Kevlar fibrils have shown a similar macrophage reaction and patchy thickening of the alveolar ducts. Both lesions again decreased after exposure, but a minimal amount of fibrosis was present 3-6 months later. A two-year inhalation study of Kevlar fibrils in rats induced exposure-related lung fibrosis (at > 25 fibres/ml) and lung neoplasms (11% at 400 fibres/ml and 6% at 100 fibres/ml in female rats; 3% at 400 fibres/ml in male rats) of an unusual type (cystic keratinizing squamous cell carcinoma). Increased mortality due to lung toxicity was observed at the highest concentration, indicating that the Maximum Tolerated Dose had been exceeded. There is considerable debate concerning the biological potential of these lesions and their relevance to humans. The full carcinogenic potential of the fibrils may not have been revealed in this study because it was terminated after 24 months.

Intratracheal instillation of a single dose of shredded Nomex paper (2.5 mg) containing fibres with diameters of 2 to 30 μm resulted in a non-specific inflammatory response. A granulomatous reaction developed two years post-exposure. Intratracheal instillation of a single dose of 25 mg Kevlar resulted in a non-specific inflammatory response which subsided within about one week. A granulomatous reaction and a minimal amount of fibrosis were observed later.

In three studies, intraperitoneal injection of Kevlar fibres (up to 25 mg/kg) resulted in a granulomatous response but no significantly increased incidence of neoplasms. It was suggested by the authors of these investigations that the lack of neoplastic response was possibly due to the agglomeration of the Kevlar fibrils in the peritoneal cavity.

There are no adequate studies in which the fibrogenic or carcinogenic potential of polyolefin fibres has been examined. A 90-day inhalation experiment in rats with respirable (46% < 1 μm) polypropylene fibres (up to 50 fibres/ml) indicated dose- and duration-dependent changes characterized by increased cellularity and bronchiolitis. No relevant data on intratracheal instillation are available. In intraperitoneal injection studies on polypropylene fibres or dust in rats, there was no significant increase in peritoneal tumours.

There are inadequate data on which to make an assessment of the *in vitro* toxicity and genotoxicity of synthetic organic fibres. For aramids, studies have shown that short and fine para-aramid fibrils have cytotoxic properties. With regard to polyolefin fibres, there is some evidence of cytotoxicity of polypropylene fibres. Mutagenicity tests on extracts of polyethylene granules gave negative results.

1.6 Effects on humans

In a cross-sectional study of 88 out of 110 workers in a PAN-based continuous filament carbon fibre production facility, there were no adverse respiratory effects, as assessed by radiographic and spirometric examination and administration of questionnaires on respiratory symptoms. In other less well documented studies, adverse effects have been reported in workers involved in the production of both carbon and polyamide fibres; data presented in the published accounts of these investigations, however, were insufficient to assess the validity of the reported associations.

1.7 Summary of evaluation

Data concerning the exposure levels of most synthetic organic fibres are limited. Those data available generally indicate low levels of exposure in the occupational environment. There is a possibility of higher exposures in future applications and uses. Virtually no data are available with respect to environmental fate, distribution, and general population exposures.

On the basis of limited toxicological data in laboratory animals, it can be concluded that there is a possibility of potential adverse health effects following inhalation exposure to these synthetic organic fibres in the occupational environment. The potential health risk associated with exposure to these synthetic organic fibres in the general environment is unknown at this time, but is likely to be very low.

2. IDENTITY, PHYSICAL AND CHEMICAL PROPERTIES, AND ANALYTICAL METHODS

2.1 Identity, physical and chemical properties

2.1.1 Carbon/graphite fibres

Carbon and graphite fibres are filamentary forms of carbon produced by high temperature transformation of one of three organic precursor materials: rayon fibres (regenerated cellulose), pitch (coal tar or petroleum residue) or polyacrylonitrile (PAN) fibres. Graphite fibres are materials characterized by a three-dimensional polycrystalline structure (Volk, 1979). The typical temperature range for graphite fibre production is 1900 to 3000 °C, while carbon fibre is manufactured at about 1200 to 1300 °C (Martin et al., 1989). In much of the literature the terms carbon and graphite fibre are used interchangeably. Trade names of carbon/graphite fibres include WCA, CCA-4, Magnamite, Thornel-T, Celion, Panex, HITEX, Fortafil, Thornel-P and Carboflex (ITA, 1985; ICF, 1986).

Rayon-based carbon fibres are very stable at high temperatures and consist of over 99% carbon. PAN-based carbon fibres have a lower carbon content (92 to 95%) and undergo further changes when heated beyond the original process temperature. The carbon content of pitch-based carbon fibres varies from 75% upward. Pitch-based fibres with a Young's modulus of over 345 GPa are almost pure (over 99.5%) carbon.

The nominal diameters of carbon fibres have been reported to range from 5 to 15 μm (Volk, 1979) and from 7 to 10 μm (Jones et al., 1982; ICF, 1986) according to end use. The distribution of fibre diameters is not available from the literature. Carbon fibres are lightweight, refractile, and have high tensile strengths, high Young's modulus (a measure of the stiffness of the material), a high degree of dimensional stability and low thermal expansion. These fibres are chemically inert and corrosion resistant, but oxidization may occur at high temperatures. They have moderate electrical and thermal conductivity (Volk, 1979). Carbon fibres are, however, vulnerable to shearing due to a decrease in interlamellar shear strength with increasing Young's modulus. The Young's modulus, electrical and thermal conductivity, and tensile strength increase with the degree of molecular orientation along the fibre axis (Volk, 1979). Strength and modulus values are greater at higher production temperatures. However, fibres with the highest Young's modulus may not have the highest tensile strength (Jones et al., 1982).

Information on the density, tensile strength and Young's modulus of carbon and graphite fibres are presented in Table 1 (Volk, 1979).

2.1.2 Aramid fibres

Aramid fibres (aromatic polyamides) are produced by a two step process involving production of the polymer followed by spinning. The polymer is typically produced by the reaction of aromatic diamines and aromatic diacid chlorides in an amide solvent. Aramid fibres are defined as fibres in which the base material is a long-chain synthetic polyamide in which at least 85% of the amide linkages are attached directly to two aromatic rings (Preston, 1979). Two types of aramid fibres are produced by the DuPont Company: Kevlar (para-aramid) and Nomex (meta-aramid), which differ primarily in the substitution positions on the aromatic ring (Preston, 1979; Reinhardt, 1980; Galli, 1981). Kevlar is made of poly(p-phenyleneterephthalamide) and Nomex is made of poly(m-phenyleneisophthalamide) (Preston, 1979). Fibres named Conex and Fenilon, which have compositions similar to that of Nomex, have also been developed in Japan (by Teijin) and the Russian Federation, respectively (Preston, 1979). Arenka, now marketed as Twaron (Holland/Germany), is a para-aramid fibre produced by Akzo. A para-aramid fibre called Arimid is produced in the Russian Federation.

Para-aramid fibres such as Kevlar/Twaron are produced as continuous filament yarn, cut (staple) fibre (38-100 mm in length), short fibre (6-12 mm in length) or pulp (2-4 mm in length), all with a nominal diameter of 12-15 μm. However, abrasive processing will produce some fibrils from para-aramid fibre. Pulp has many fine-curled, tangled fibrils within the respirable size range attached to the surface of the core fibre, and some of these fibrils will break off and become airborne during manufacture or use. These fibrils have a ribbon-like morphology and may have widths less than 1 μm. Figure 1 shows a scanning electron microscopy of ultrafine Kevlar fibrils peeling off a 12-μm Kevlar fibre (Warheit et al., 1992).

Meta-aramid fibres such as Nomex/Conex are manufactured as continuous filament, staple fibre and short fibre, all with a

Table 1. Physical and chemical properties of carbon/graphite fibres[a]

Fibre	Density (g/cm^3)	Tensile strength (MPa)	Young's modulus (GPa)
Rayon-based carbon fibres (low modulus)	1.43-1.7	345-690	21-55
Rayon-based carbon fibres (high modulus)	1.65-1.82	-	345-517
PAN-based carbon fibres	1.7-1.8	2400-2750	193-241
Pitch-based carbon fibres (filament yarn)	2.0	2000	345

[a] From: Volk (1979)

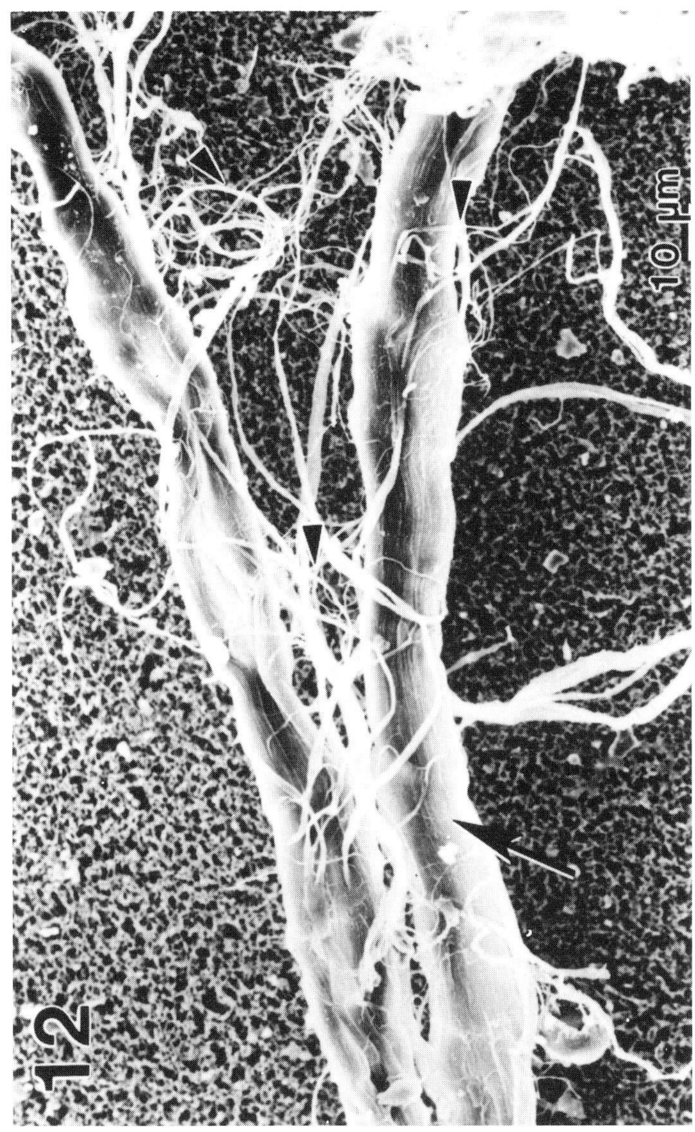

Fig. 1. Scanning electron micrograph of ultrafine Kevlar fibrils (arrowheads) caught on an aerosol filter used for fibre counting (Warheit et al, 1992; with the permission of Academic Press, Orlando, Florida, USA)

nominal diameter of approximately 12 μm. Unlike para-aramid fibres, meta-aramids have no fibrillar substructure and do not tend to produce smaller diameter fibrils upon abrasion (ICF, 1986).

Generally, aramid fibres have medium to very high tensile strength, medium to low elongation-to-break, and moderate to very high modulus (Preston, 1979; Brown & Power, 1982). Meta-aramid fibre of low orientation has a density of approximately 1.35 g/cm^3 and the hot-drawn fibre has a density of approximately 1.38 g/cm^3. Para-aramid fibres of relatively high crystallinity have a density of 1.44 g/cm^3. The volume resistivities and dielectric strengths of these fibres are high, even at elevated temperatures (Preston, 1979). Aramid fibres are heat resistant, with mechanical properties being retained at temperatures up to 300-350 °C (Preston, 1979). Whole aramid fibres are generally resistant to chemicals, with the exception of strong mineral acids and bases (Preston, 1979; Galli, 1981; Chiao & Chiao, 1982). The strength to weight ratio of Kevlar is high; on a weight basis, it is five times as strong as steel, ten times as strong as aluminum and up to three times as strong as E-glass. Aramid fibres have excellent toughness, and withstand continuous heat at temperatures in the 160-205 °C range. Aramids will not melt or support combustion, and carbonization will not be appreciable under 400 °C. Para-aramid fibres have a small negative coefficient of longitudinal thermal expansion, similar to graphite, making them suitable for joint usage in composites (Galli, 1981; Hanson, 1980). Some physical and chemical properties of aramid fibres are presented in Table 2 (Hodgson, 1989).

2.1.3 Polyolefin fibres

Polyolefin fibres are long-chain polymers composed (at least 85% by weight) of ethylene, propylene or other olefin units (Buchanan, 1984). Although many different types of polyolefins are produced, this term is commonly used only for hydrocarbon polyolefins, polyethylene and polypropylene (Hartshorne & Laing, 1984). These are the fibres considered in this monograph. Approximately 95% of all polyolefin fibres are polypropylene (Ahmed, 1982).

Structurally, the high-density polyethylenes consist of linear methylene chains usually terminated by vinyl but occasionally by methyl groups. The low-density analogue also contains methylene chains, but has frequent branches. Chains are usually terminated

Table 2. Physical and chemical properties of aramid fibres[a]

Fibre type	Density (g/cm^3)	Tensile strength (MPa)	Specific strength (MPa/d)	Young's modulus (GPa)	Flammability (LOI) at room temperature	Approximate thermal degradation (°C)
Kevlar/Twaron	1.44-1.45	2790-3000	2000	120-124	24.5	> 400
Nomex	1.38	-	-	-	26	> 370

[a] From: Hodgson (1989)

by methyl groups. An intermediate density polyethylene is also available, together with blends of polyethylene with polypropylene or polyisobutene for use mainly in ropes or twines. Trade names of polyolefin fibres include BDH Low Density, Courlene C3, Courlene X3, Courlene Y3, Downspun 82, Dyneema, Fibrite MF, Meraklon BCF, Meraklon DO, Novatron, Polyolefine, Sanylene, and Spratra (Hartshorne & Laing, 1984).

Polyolefin fibres and products are produced in eight different forms (ICF, 1986):

1) monofilament yarn typically having a diameter of 150 μm or more;

2) multifilament yarn similar in structure to monofilament yarn but for which the diameter of individual filaments is generally in the range of 5 to 20 μm has also been reported;

3) staple fibre, which is multifilament yarn cut into varying lengths of up to a few millimeters and which may be used in either woven or discontinuous form;

4) tape and fibrillated film yarn with a large rectangular diameter resembling a ribbon; the thickness typically varies from 2.5 to 12.7 μm for tape and 2.5 to 6.4 μm for fibrillated film yarn;

5) spun-bonded fabric, i.e. non-woven fibrous structures produced in the form of flat fabric in one continuous process directly from the molten resin;

6) synthetic pulp - a relatively new class of discontinuous fibres with diameters ranging from 5 to 40 μm and maximum lengths of 2.5 to 3 mm;

7) meltblown or microfibre - a relatively new type of polyolefin fibre produced by the meltblown process for which the diameter typically ranges from 0.1 to 2 μm and the length is a few centimetres;

8) high strength, high modulus, highly oriented, high molecular weight polyethylene fibres formed by spinning, drawing a partially polymerized gel.

The specific gravities of polyethylene and polypropylene are low and they are highly hydrophobic; properties of these materials in wet conditions are, therefore, similar to those measured at standard temperature and humidity (21 °C; 65% relative humidity). The polyolefins are unaffected by a wide variety of inorganic acids and bases and organic solvents at room temperature. The melting point of low-density polyethylenes is less than 116 °C and of high-density ones is above 131 °C. Medium-density polyethylene is reported to melt at 125 °C, whereas polypropylenes melt at between 167 and 179 °C (Hartshorne & Laing, 1984). Some physical and chemical properties of the polyolefins are presented in Table 3 (Hodgson, 1989).

2.2 Production methods

2.2.1 Carbon/graphite fibres

In general, the process of carbon and graphite fibre production involves three distinct steps: preparation and heat treating, carbonization, and optional high-temperature annealing (Volk, 1979). Initially, the precursors (PAN, pitch or rayon) are oxidatively stabilized and dehydrogenated at moderate temperatures (200–300 °C). For pitch-based fibres, the commercial coal or petroleum pitch is converted through heat treatment into a mesophase or liquid crystal state. The fibre is then carbonized at temperatures of 750–1375 °C in a non-oxidizing atmosphere, and production may involve a secondary heating phase, known as graphitization, at approximately 1400 °C in an inert atmosphere.

Mesophase pitch-based fibres having a Young's modulus exceeding 690 GPa, made by heating without stretching to 3000 °C, are the most graphitic in nature (Volk, 1979). To produce high modulus fibres, yarn is stretched during the last heat treatment. Depending on the configuration of the final product (e.g., chopped fibre, continuous strand, felt or fabric) and intended use, the fibre may be treated with a sizing material to improve its handling characteristics and compatibility with various matrices. The formulation of sizing compounds (protective coatings) is considered proprietary; it has been reported, however, that they are typically compounds with epoxy-based functionalities, resins, epoxides and aqueous systems (ICF, 1986). Fibres may be chopped to lengths in the range of 3 to 25 mm for use in reinforced composite materials or milled down to 200 μm for special conductive applications. For composites, the fibrous material is embedded in a matrix, such as an epoxy resin, to form strong, lightweight engineering materials (Martin et al., 1989).

2.2.2 Aramid fibres

Aramid polymers are made by solution polymerization. This involves low temperature polycondensation of diacid chlorides and

Table 3. Physical and chemical properties of polyolefin fibres[a]

Polymer	Fibre type	Density (g/cm^3)	Young modulus (GPa)	Tensile strength (MPa)	Elongation to break (%)
Low-density polyethylene	monofilament	0.92	0.8-1.0	-	-
High-density polyethylene	monofilament	0.95-0.96	1.7-4.2	-	-
Polyethylene	high density filament	0.96	1.7-4.2	290-570	10-45
Gel-spun polyethylene	filament	-	44-77	2580-5500	-
Polypropylene	staple and tow	0.90-0.96	0.28-3.3	-	-
	monofilament	0.90-0.91	1.6-4.8	-	-
	multifilament	0.90-0.91	1.2-3.2	-	-
Polypropylene	filament	0.91	1.6-4.8	270-540	14-30
	high tenacity	-	-	811	-

[a] From: Hodgson (1989)

diamines in amide solvents. Meta-aramid fibre is spun from the polymerization solution of dimethylacetamide after neutralization. Para-aramid polymer must first be neutralized and isolated from the polymerization solution; it is then re-dissolved in a spinning solution of concentrated sulfuric acid. This liquid crystalline solution is then extruded through a spinneret, and the acid is extracted and neutralized to form a highly-oriented fibre.

2.2.3 Polyolefin fibres

Low-density polyethylene is manufactured by a high-temperature and high-pressure polymerization process, whereas the high-density form is made at low temperatures and pressures using a more efficient catalyst (e.g., the Marlex process). Intermediate density polyethylene and polypropylene are produced with a catalyst by the Ziegler process (Hartshorne & Laing, 1984).

There are several processes for the production of continuous and discontinuous polyolefin fibres (ICF, 1986). Mono-filament and multi-filament yarns are formed by the continuous extrusion of the molten polymer through a spinneret, solidification by heat transfer, and winding onto packages. Processing may include drawing of the fibre to up to six times its original length, as well as heat treatment to relieve thermal stress. Staple fibre is cut from multi-filament yarn. Yarns are also formed by splitting a highly oriented film. Flash spun fibres are formed by dissolving polyolefin polymer under pressure and flashing from a spinneret to form fibres. If done at high pressure the fibres are much shorter and can be used as a synthetic pulp. Melt blown fibres are formed by drawing a spun fibre in high pressure steam or air.

2.3 Sampling and analytical methods

Sampling and analytical methods for organic fibres include the measurement of total airborne or respirable mass concentration and the determination of airborne fibre number concentrations by phase contrast optical microscopy (PCOM). Sampling methods used for organic fibres are similar to those used for inorganic fibres such as asbestos or man-made mineral fibres. These methods typically involve drawing a measured volume of air through a filter mounted in a holder that is located in the breathing zone of the subject. For measurement of mass concentrations, either polyvinyl chloride or glass fibre filters are normally used. The filters are stabilized in air and weighed against control filters, both before and after sampling, to permit correction of weight changes caused by varying humidity. Cellulose ester membrane filters are usually used for assessing fibre number concentrations. In this case, the filter is made optically transparent with one of several clearing agents (e.g., triacetin, acetone or ethylene glycol monomethyl ether), and the fibres present in random areas are counted and classified using PCOM (WHO, 1985; NIOSH, 1989 [Method 7400]). Different fibre counting rules have been used in various countries and these may give somewhat different values.

Although the basic methods for the determination of total airborne mass and fibre number concentrations are similar in most countries, differences in the sampling procedure, the filter size and type, the clearing agent and the microscope used, and subjective errors in sampling and counting all contribute to variations in results. Specific reference methods for the determination of organic fibres have not been developed. However, the methods mentioned above have been used for routine industrial hygiene monitoring. A WHO project is in progress to develop a reference method for the measurement of health risk-related fibres in workplace air.

Several potential problems exist with respect to use of PCOM sampling and analytical methods for organic fibres. Firstly, these fibres have significant electrostatic charges that could affect sampling efficiency. Secondly, some of these fibres may be soluble in the microscope slide mounting medium, and their visibility in PCOM has not been evaluated. Lastly, PCOM methods lack specificity for counting organic fibres.

The improved resolution of electron microscopy and the identification capacity particularly of the analytical transmission electron microscope (TEM), with selected area electron diffraction (SAED) and energy dispersive X-ray analysis (EDXA), make these methods particularly suitable for more complete characterization of the fibre size distribution and analysis of fibres of small diameter (NIOSH, 1989 [Method 7402]). However, these methods have so far rarely been used for analyses of organic fibres.

For analysis by scanning electron microscopy (SEM), fibres collected on polycarbonate filters can be examined directly. This avoids the need for transfer techniques that may affect the fibre size distribution. For TEM, direct transfer preparation techniques involving carbon coating of particles on the surface of a polycarbonate or membrane filter and indirect transfer methods in which attempts have been made to retain the fibre size distribution are the most widely accepted.

3. SOURCES OF HUMAN AND ENVIRONMENTAL EXPOSURE

3.1 Production

3.1.1 Carbon/graphite fibres

The principal producers of carbon and graphite fibres are Japan, the USA and the United Kingdom, although these materials are also manufactured in France, Germany and Israel (ITA, 1985; ICF, 1986). The raw material most commonly used is polyacrylonitrile (PAN); in Japan, the world's largest producer of PAN-based fibres, the precursor material is obtained from the excess acrylic fibre capacity of the Japanese petrochemical industry. The USA is the main world producer of carbon/graphite fibres from rayon and pitch (ICF, 1986).

World production capacity was estimated to be 4173 tonnes in 1984. Based on information provided by manufacturers in the USA, estimated world consumption for 1984 and 1987 was 889 tonnes and 2046 tonnes, respectively. Similar estimates based on information provided by companies in Japan were 1068 tonnes and 1756 tonnes, respectively. The USA and Japan are the principal consumers, followed by western Europe (ITA, 1985). More recently, it has been reported that the total world capacity for carbon fibres is 10 000 tonnes per annum, although the author noted that it was difficult to assess what this means in terms of real annual output (Hodgson, 1989).

The estimated consumption of rayon-based carbon fibres from 1970 to 1976 was 50 to 125 tonnes (Volk, 1979). It has also been estimated that the combined production of PAN-based carbon fibres in the three principal producing countries (USA, Japan and the United Kingdom) rose from 10 tonnes in 1970 to 250 tonnes in 1976 (Volk, 1979).

3.1.2 Aramid fibres

Kevlar and Nomex which are both produced by DuPont have been sold commercially since 1970 and 1965, respectively (Reinhardt, 1980). The production capacity for Kevlar in 1978 was reported to be approximately 3400 tonnes (Galli, 1981). In 1975, the production capacity for Nomex was expanded from 4500 tonnes to more than 9100 tonnes (Preston, 1979).

More recently, it has been reported that the production capacity per year of Kevlar (USA) and Twaron (Holland/Germany) is 20 000 and 5000 tonnes, respectively (Hodgson, 1989). Plants in Northern Ireland and Japan with annual production capacities of about 5000 tonnes per year each have been brought on line since then.

3.1.3 Polyolefin fibres

Production of polyolefin fibres has been increasing owing to greater demand for continuous filament yarn, monofilaments and staple fibres (Buchanan, 1984). Production and consumption in the USA in 1983 were approximately 182 500 and 175 000 tonnes, respectively (SRI International, 1985). The annual polyolefin fibre production capacities of producers in the USA have been reported to range from approximately 4500 kg to more than 134 000 kg (SRI International, 1987).

3.2 Uses

Carbon and aramid fibres are used principally in advanced composite materials to improve strength, stiffness, durability, electrical conductivity or heat resistance. Since these fibres improve properties such as these without adding much weight, they are used principally in the aerospace industry, for military reasons and in sports equipment manufacture (ITA, 1985). Polyolefin fibres are typically used in textile applications, although gel spun fibres are used for high tensile strength, but low temperature, applications.

3.2.1 Carbon/graphite fibres

In addition to the general categories mentioned above, carbon fibre in felt form is used in high temperature insulation, principally in inert atmospheres due to the tendency of the material to oxidize.

Rayon-based carbon fibres are used primarily in aerospace applications, such as phenolic-impregnated heat shields and carbon-carbon composites for missile parts and aircraft brakes (Volk, 1979). PAN-based carbon fibres are used in structural applications in the aerospace industry, sports goods (golf clubs, fishing rods, tennis rackets, bows and arrows, skis, sailboat masts and spars), reinforcement for moulded plastics, prostheses, artificial joints and dental bridges (Volk, 1979; Bjork et al., 1986).

Pitch-based carbon fibres are used in applications to increase electrical conduction and resistance to heat distortion and to improve wear and stiffness (e.g., as veil mat for sheet moulding, milled mat in injection moulding) (Volk, 1979).

3.2.2 Aramid fibres

Properties of aramid fibres, such as heat and flame resistance, dimensional stability, ultra-high strength and high modulus, electrical resistivity, chemical inertness and permselective properties, make them useful in many applications (Preston, 1979).

Para-aramids are used principally as a strengthening and reinforcing material in composite structures due to their low density, high specific strength and stiffness, greater vibration damping and better resistance to crack propagation and fatigue than those of inorganic fibrous materials. Para-aramids are used primarily for tyre cords, protective clothing, industrial fabrics, high performance (sports and aerospace) composites, high-strength ropes, cables, friction materials and gaskets (ILO, 1989).

Meta-aramids are primarily used for their heat and corrosion resistance (Brown & Power, 1982). The paper form is used as electrical insulation in motors and transformers. Staple and continuous filaments are used in protective clothing and coated fabrics, and industrial filter bags for hot gas emissions.

Aramid fibres are sometimes combined in various applications with other fibrous materials such as carbon and graphite to reduce costs and increase impact strength (Delmonte, 1981).

3.2.3 Polyolefin fibres

Apart from their traditional use in carpet backings, polyolefin fibres are being used increasingly in other household furnishings such as upholstery, bedding, curtains, wall-covering materials and carpet pile (Hartshorne & Laing, 1984; Buchanan, 1984). The largest use of polyolefin fabric in clothing is in disposable nappies (diapers) and athletic socks.

Polyolefin fibres are also used in ropes, cordage and twine, webbing, synthetic turf, agricultural fabrics, commercial fishing lines and nets, sewing thread and book binders (Buchanan, 1984). They are also used in the medical field in sutures and as matrices for tissue ingrowth and anchoring (Hoffman, 1977).

Microfibres are used in a variety of applications including filtration, medical/surgical fabrics, sanitary products, sorbents, wipes, apparel insulation, protective clothing and battery separators.

3.3 Emissions into the environment

3.3.1 Fibre emissions

Carbon fibres may be released into the environment during production, processing or combustion of materials made of or containing carbon composites. These may arise during the incineration of waste material containing carbon fibres and aircraft fires. Light microscopy analysis of emissions from several key operations in the manufacture and processing of carbon fibres (weaving, prepregging, machining and incineration of composites) showed release rates (fibre mass released per unit of material processed) varying over several orders of magnitude, with weaving and incineration being the greatest (Gieseke et al., 1984). Fibre diameters were found to be reduced only during incineration.

Based on laboratory studies in which composites containing carbon fibres were burnt, it was concluded that the rate of emissions and lengths of fibres released in municipal incineration are dependent on air flow rates and mechanical agitation within the combustion chamber. The diameters of emitted fibres were dependant upon burning rate, temperature and oxygen levels, and on fibre residence time in the incinerator. It was estimated that a typical emission rate of carbon fibres from the burning zone in a municipal incinerator would probably be in the order of 1% of the composite being burnt. It was also concluded that composites made with epoxy binder materials release more fibres during incineration than those made with phenolic binder materials (Gieseke et al., 1984).

In studies conducted to simulate municipal incineration of refuse from composites containing carbon fibres (combustion rate of 34 kg/h, flame temperature of 925 °C, 450 kg Celion unidirectional laminate fed over a 1-min period), stack emission rates were 0.840 to 0.998 mg/dscm carbon fibre. Emission rates decreased with time (0 to 60 min) following introduction of the refuse from the carbon fibre composite (Henry et al., 1982; Gieseke et al., 1984).

In laboratory and field tests conducted by the United States National Aeronautics and Space Administration, it was estimated that approximately 38.4-49.5% of the fibres released during the burning of composite materials containing PAN-based carbon/graphite fibres in aircraft (spoilers, cockpits, and various structural elements) were respirable (i.e. aspect ratio of $\geq$ 3:1; < 80 μm in length and < 3 μm in diameter). Based on analysis by optical microscopy, the average length of emitted fibres was 30 μm (Sussholz, 1980).

Based on data collected in his previous studies, the same investigator estimated airborne fibre concentrations during the burning of carbon fibre composites. For a fire resulting from an aircraft crash, it was estimated that 1% of the carbon fibre would be released, constituting about 5×10^{11} respirable fibres per kg. Based on these data, it was further estimated that the airborne concentration of fibres in the densest part of the smoke would be 5 fibres/ml.

The by-products of carbonization that are captured following incineration of fibres include hydrogen cyanide from PAN-based fibres and polynuclear aromatic compounds from pitch-based fibres. By-products of graphitization include ammonia, cyanide and hydrocarbons; these materials are decomposed during the heating process (ICF, 1986).

3.3.2 Decomposition products

In studies conducted to simulate municipal incineration of refuse from composites containing carbon fibres (combustion rate of 34 kg/h, flame temperature of 925 °C, 450 kg Celion unidirectional laminate fed over 1-min period), observation by light microscopy indicated that fibres are oxidized during incineration and that fibre diameters are reduced. It was also observed that as oxidation proceeds, there is notching and subsequent breaking of the fibres. Thus, oxidized fibres that have small diameters are often relatively short (Henry et al., 1982).

The burning or pyrolysis of organic fibres produces a wide variety and quantity of emitted gases depending on the numerous variables of the decomposition/oxidation. Simultaneously burning materials, temperature, heating rate, burning time, humidity and oxygen availability are just a few of the important variables. Consequently, the data from a given experiment are uniquely dependent on the conditions of that experiment.

Pyrolysis decomposition products of carbon and Kevlar fibres have been identified by thermogravimetric analysis/gas chromatography/mass spectrometry (TGA/GC/MS) (Razinet et al., 1976). The products identified in the off-gasses of pyrolysis of 11 different types of carbon fibre at 1000 °C included carbon dioxide (32-100%), methane (2-23%), propane (0.5-19%), propene (2-20%), benzene (7-14.4%), toluene (2.0-8.3%), dimethylbenzene (2.9%), naphthalene (2.9%) and methylnaphthalene (2.8%). The carbon dioxide content of the off-gasses (expressed as a percentage of the weight of the sample used in the decomposition) ranged from 0.02 to 2.34%.

The products identified after fast pyrolysis of Kevlar (650 °C) were carbon monoxide (63%), benzene (35%) and toluene (2%). Following slow pyrolysis (at 650 °C for 15 min), many compounds, including nitrogen, carbon monoxide, carbon dioxide, methane, ethylene, propene, benzene, toluene, benzonitrile, aniline, methylaniline, phenylacetonitrile, phthalonitrile, phenyldiamine, biphenyl, benzimidazole, fluorene and benzanilide, were identified.

4. ENVIRONMENTAL LEVELS AND HUMAN EXPOSURE

4.1 Occupational environment

4.1.1 Carbon/graphite fibres

4.1.1.1 Production

Carbon fibres are released in the workplace during operations such as fibre forming, winding, chopping and composite formation. Analysis by light microscopy of emissions in the immediate vicinity of each of several key operations in the manufacture and processing of carbon fibre (including weaving, handling, prepregging, and machining of composites) demonstrated that release rates (fibre mass released per unit of material processed) were greatest for weaving (Gieseke et al., 1984).

The data on aerosol exposures to carbon fibres in the occupational environment are presented in Table 4. Due to the lack of available data and limited information on methods for sampling and analysis, only tentative conclusions concerning airborne particle or fibre concentrations can be drawn.

4.1.1.2 Processing of composites

Some data are available concerning airborne concentrations and fibre dimensions resulting from the machining of carbon fibre composites. Following drilling, an analysis of the settled dust was conducted using PCOM and scanning electron microscopy (SEM). Fibres 50 to 100 μm in length with diameters 6 to 8 μm were observed and found to have dimensions similar to those of fibres in the composite material. Following sawing, settled dust samples were also collected and analysed, and this revealed longitudinal breaking of fibres. Some fibres were observed to have fibre diameters less than those in the parent material; fibres were also shorter in length than those in the dust produced by drilling (Wagman et al., 1979).

Limited data are available concerning the preparation and machining of carbon fibre composites. PCOM analyses of five samples from one facility yielded concentrations ranging from < 0.001 fibres/ml to approximately 0.01 fibres/ml. The average length of fibres in these samples ranged from approximately

Table 4. Airborne concentrations of carbon fibre in the workplace[a]

Sampling and analysis	Results	Reference
Light microscopic analysis of unspecified number of samples (possibly one each) of "outside emissions" at prepregging machining and weaving operations in PAN-based production facility in the USA	0.0002 fibres/ml in prepregging, mean fibre diameter 6.1 μm and mean fibre length 213 μm; 0.0009 fibres/ml in shuttle loom weaving, mean fibre diameter 6.7 μm and mean fibre length 749 μm; 0.003 fibres/ml in rapier weaving, mean diameter 3.9 μm and mean fibre length 706 μm; 0.01 and 0.001 fibres/ml in machining, mean fibre diameters 6.2 and 6.6 μm, and mean fibre lengths 32.8 and 30.9 μm, respectively	Henry et al. (1982)
Light microscopic analysis of 4 samples of either "outside emissions" or worker exposure (not clearly specified) in winding operations in PAN-based production facility in the USA	mean concentration of 0.0005 fibres/ml (range, 0.00002-0.006 fibres/ml); average fibre diameter 6.5 μm and average fibre length 48-60 μm	Henry et al. (1982)
Gravimetric analysis of 38 samples in pitch-based continuous filament production facility	mean concentration of 0.08-0.39 mg/m^3 total dust with 40% respirable; concentrations highest in the laboratory owing to cutting, grinding and milling of carbon fibre; reinforced resins (mean, 0.39 mg/m^3), concentration twice as high in winding (mean, 0.19 mg/m^3) as in production (mean, 0.08 mg/m^3)	Jones et al. (1982)

Table 4 (contd).

Sampling and analysis	Results	Reference
Light microscopic analysis of emissions at several locations each in various phases of carbon fibre production	0.003-0.029 fibres/ml in prepegging, mean fibre diameter 5.5-6 μm and mean fibre length 915-2399 μm; 0.001-0.05 fibres/ml in weaving, mean fibre diameter 5.5-7.8 μm and mean fibre length 945-2342 μm	Gieseke et al. (1984)
Gravimetric analysis of 6 breathing zone samples in PAN-based production facility in the USA	0.10-0.80 mg/m^3; concentrations much higher for chopping and winding operators (0.54-0.8 mg/m^3) than for line operators (0.1-0.12 mg/m^3)	Gilliam (1986)[b]
PCOM analysis (NIOSH 239) of all areas of a pitch-based production facility in the USA where exposure was deemed likely (1985 to May 1986)	0.11-0.27 fibres/ml	Familia (1986)[c]

[a] This study was designed to estimate emissions from carbon fibre processing. The Task Group noted that these data may be of limited value for predicting occupational exposure levels and airborne fibre characteristics.

[b] Gilliam, H.K. Great Lakes Carbon Corporation, Rockwood, TB 37854-0810. Letter providing monitoring data and product literature to James C. Chang, ICF Inc., Washington, DC, April 28, 1986.

[c] Familia, R.A. Union Carbide/Amoco. Greenville, SC 29602. Letter providing sampling data to James Chang, ICF Inc., Washington, DC, (1986).

31 μm to 749 μm and average diameter from 3.9 μm to 6.7 μm (Henry et al., 1982).

The mean airborne concentration determined by PCOM and SEM was 0.03 fibres/ml in fourteen samples taken during the machining of fibre composites at the Air Force Aerospace Materials Research Laboratory. Particles with an aspect ratio of > 3 constituted less than 0.1% of the released material. The diameters (> 7 μm) of the airborne fibres were similar to those in the composites (Lurker & Speer, 1984).

Airborne particulate samples were collected during trimming operations on a wing panel composed of carbon fibre composites. Based on analysis by PCOM, less than 8% of the airborne particulates were fibrous. The diameters of at least 80% of the fibres were approximately 7 μm, i.e. similar to those in the composites (Dahlquist, 1984).

The physical, morphological and chemical characteristics of machined graphite or fibrous glass composites have been determined. Bulk and fractionated samples were examined by light and electron microscopy and analysed chemically by GC/MS (Boatman et al., 1988). The test materials were designated as "graphite-p", two graphite materials manufactured from PAN, one pitch-based graphite and PAN-graphite/Kevlar. The machining operations were spindle shaping (at 3450 or 10 000 rpm), hand routing (at 23 000 rpm) and use of a saber saw. Dusts were collected with a high efficiency vacuum cleaner and resuspended in a dust generator to create an aerosol in which 90% of particles were < 10 μm in aerodynamic diameter. The relative proportion of respirable to total mass of bulk samples was < 3%; aerodynamic diameters of fractionated samples collected at the tool face ranged from 0.8 to 2.0 μm. Particles in bulk samples ranged from 7 to 11 μm in diameter, and mean aspect ratios ranged from 4:1 to 8:1 (26:1 for the fibrous glass composite). The mean diameter of fractionated particles collected at the tool face was 2.7 μm, with 74% being less than 3.0 μm in diameter. There was no evidence of longitudinal splitting of fibres, and volatilization of chemicals from machined composites was low. The authors reported that the respirable fractions in dusts collected at the tool face were only 2 to 3% by weight. However, since large particles that would settle rapidly and not normally reach the breathing zone were included in samples taken at the tool face, these values are not indicative of personal exposure.

4.1.2 Aramid fibres

4.1.2.1 Production

Available data on airborne concentrations of polyamide fibres in the occupational environment are restricted to brief summaries of unpublished data collected within the industry. In many cases, methods of sampling and analysis have not been described.

Verwijst (undated report) described exposure monitoring during para-aramid fibre and pulp manufacturing and during laboratory operations. Air concentrations ranged from 0.01 to 0.1 fibrils/ml, with the highest values being for pulping. Verwijst also noted relatively high exposure (0.9 fibrils/ml) during water jet cutting of composites.

Since initiation of Kevlar para-aramid fibre production (in about 1971), plant-air and employee exposures have been measured by the same phase contrast microscopy techniques used for asbestos (PCAM 239 before about 1982 and NIOSH 7400 "A" rules) (Merriman, 1992). These data are summarized in Table 5. For continuous filament yarn handling, plant exposures are extremely low (0.02 fibres/ml maximum) (Reinhardt, 1980). Cutting of staple and floc fibre produced levels of 0.2 fibres/ml or less with a single peak measurement of 0.4 fibres/ml. Pulp drying and packaging operations led to maximum concentrations of 0.09 fibres/ml.

Airborne respirable concentrations of Nomex/Conex have been reported to be less than the limit of detection i.e. 0.01 fibres/ml (Reinhardt, 1980; ILO, 1989).

4.1.2.2 End-use processing and processing of composites

Para-aramid end-use plant monitoring data using PCOM are summarized in Table 5 for brake pad production, gasket composite fabrication, and staple yarn spinning processes (Merriman, 1992). No exposures exceeded 0.19 fibres/ml for brake pad manufacturing, where dry para-aramid pulp was mixed with powdered fillers and resin, pressed, cured, ground and drilled. Average fibre exposures were less than 0.1 fibres/ml.

In gasket sheet and gasket manufacturing, para-aramid pulp is mixed with fillers and solvated rubber cement, rolled into sheets and die-cut into smaller pieces that may be finished by sanding the edges. A total of 64 samples in four plants gave no personal exposures greater than 0.15 fibres/ml and no area concentrations greater than 0.25 fibres/ml. Mean exposures were less than 0.1 fibres/ml for all operations. In a single test where heat-aged gaskets were scraped from flanges, short-term exposure concentrations were less than 0.2 fibres/ml.

Table 5. Airborne fibre concentrations in workplaces
handling para-aramid fibre pulp[a]

Manufacturing industry	Operations	Samples	Mean (fibres/ml)	Maximum (fibres/ml)
Brake pads	mixing	20	0.07	0.15
	preforming	17	0.08	0.19
	grinding/drilling	8	0.04	0.08
	finishing/inspecting	3	0.05	0.10
Gaskets	mixing	30	0.05	0.15
	calendering	1	0.02	0.02
	grinding	5	0.08	0.25
	cutting	15	0.02	0.07
Composite	sanding/trimming	5	0.08	0.25
	water jet cutting	1	0.03	2.91
Staple yarn	grinding	5	0.18	0.28
	carding	16	0.39	0.79
	drawing	4	0.32	0.87
	roving	6	0.33	0.72
	spinning	15	0.18	0.57
	twisting/winding	13	0.55	2.03
	finishing	2	0.30	0.48
	weaving	6	0.35	0.58

From: Merriman (1992)

Machining of para-aramid fabric-reinforced organic matrix composites produced very low exposures; most were less than 0.1 fibres/ml although one exposure reached 0.25 fibres/ml during grinding. Although operator exposure during water-jet cutting was only 0.03 fibres/ml, the cutting sludge in a single sample was highly enriched with respirable fibrils and produced much higher levels (2.9 fibres/ml).

The end-use most likely to produce significant para-aramid fibril exposure levels has been found to be staple fibre carding and subsequent processing into yarn. Carding is highly abrasive and the fibrils produced are immediately entrained in the high air flows created by the spinning cylinders. Monitoring of operators in six yarn-spinning mills (67 personal samples) gave average exposures ranging from 0.18 to 0.55 fibres/ml, with one operation reaching a maximum of 2.03 fibres/ml.

Kauffer et al. (undated report) characterized airborne fibre concentrations and size distributions during the monitoring of machining of carbon fibre- and aramid-based composites in industry and the laboratory. Concentrations encountered were typically well below 1 fibril/ml, as determined by optical microscopy. SEM showed mean lengths from 1.9 to 4.3 μm, and mean length/diameter ratios values varied from 4.4 to 8.8. The authors concluded that most of the respirable material consisted of resin debris.

Particle and gaseous emissions during laser cutting of aramid fibre-reinforced epoxy plastics have been studied (Busch et al., 1989). The mass-median aerodynamic diameter of particles generated was 0.21 μm, but the concentration of dust and the fibre content of the dust were not reported. GC/MS analyses of samples on charcoal and silica tubes demonstrated the following release of gases per gram of material pyrolized during cutting: 5.4 mg benzene, 2.7 mg toluene, 0.45 mg phenylacetylene, 1.4 mg benzonitrile, 1.0 mg styrene, 0.55 mg ethylbenzene, 0.15 mg m-xylene and p-xylene, 0.04 mg o-xylene, 0.28 mg indene, 0.16 mg benzofurane, 0.15 mg naphthalene, and 0.73 mg phenol.

Limited personal exposure monitoring was conducted during laser cutting of Kevlar-reinforced epoxy matrix (Moss & Seitz, 1990). An air sample collected within a few feet of the cutting operation revealed few fibres (0.15-0.25 μm in diameter and < 10 μm in length) in TEM analyses. In addition to fibre measurements, hydrogen cyanide concentrations in the cutting area ranged from 0.03 to 0.08 mg/m^3 with a TWA of 0.05 mg/m^3. Carbon monoxide concentrations ranged from 10 to 35 ppm and nitrogen dioxide concentrations were < 0.5 to 5 ppm.

4.1.3 Polyolefin

4.1.3.1 Production

Limited air monitoring data in a facility producing polypropylene fibres have been reported (Hesterberg et al., 1991). Samples for PCOM analyses were collected in the breathing zone of workers, and fibres were counted using NIOSH method 7400 ("B" rules). Gravimetric measurements of total dust exposures were also made. Airborne levels of fibres longer than 5 μm ranged from none detected to a peak of 0.5 fibres/ml, with most values being less than 0.1 fibres/ml. The total dust exposures ranged from below detection limits to 0.7 mg/m^3, most values being less than 0.25 mg/m^3. SEM analyses showed airborne fibre diameters to range from 0.25 to 3.5 μm (mean = 1.97) and lengths to range from 1.68 to 69 μm (mean = 29.4 μm).

4.2 General environment

Henry et al. (1982) reported the results of limited sampling and analysis by light microscopy of the concentration of carbon fibres in ambient air in the vicinity of carbon fibre facilities. The concentration of fibres in ambient air downstream of the baghouse for rapier weaving was reported to be 0.0003 fibres/ml (Henry et al., 1982). The average length of fibres was 706 μm and the average width 3.9 μm.

The release of carbon fibres at the crash and burn site of two military aircrafts has been described (Formisano, 1989; Mahar, 1990). Both aircrafts were manufactured using carbon fibre composites and the air sampling was conducted several days after the crash. At the first site 21 area samples and 11 personal samples were taken over four days. The mean durations of the area and personal samples, respectively, were 168 ± 86 min and 48 ± 46 min at 2 litres/min. Twelve of the 21 area samples contained low fibre concentrations and were considered to be estimates. The mean fibre concentration of the remaining air samples was 0.29 ± 0.41 (range 0.015–1.060) fibres/ml. In the case of the 11 personal samples, 4 were estimates and the mean concentration of the others was 3.40 ± 2.52 (range 0.063–6.998) fibres/ml.

A mixture of floor-wax and water was poured over the aircraft wreckage in both cases to suppress the dust. The personal air samples were collected while a variety of tasks commonly performed at a crash site were performed. The author reported that fibres may have been lost by electrostatic binding to the sampling cassette. While a few interfering fibres were present, the straight fibres with clean edges were recognized by the analyst as man-made fibres. Air sampling at the second crash site produced barely detectable levels, and the highest concentration determined by PCOM was 0.58 fibres/ml.

5. DEPOSITION, CLEARANCE, RETENTION, DURABILITY AND TRANSLOCATION

5.1 Introduction

It is considered that the potential respiratory health effects related to exposure to fibre aerosols are a function of the internal dose to the target tissue, which is determined by airborne concentrations, patterns of exposure, fibre shape, diameter and length (which affect lung deposition and clearance) and biopersistence. The potential responses to fibres, once they are deposited in the lungs, are a function of their individual characteristics. In inhalation studies in rodents, fibres with dimensions similar to those that humans can inhale should be used, provided that a high proportion is within the range respirable for the rat. In addition, complete characterization (e.g., dimensions, number, mass and aerodynamic diameter) of the rat-respirable fibre aerosol and retained dose is essential. Methods for aerosol generation should insure that fibre lengths are preserved.

The following general principles have been derived principally on the basis of results of studies with particulates, man-made mineral fibres and asbestos. It should be recognized, however, that these parameters have not been evaluated in detail specifically for the synthetic organic fibres.

Because of the tendency of fibres to align parallel to the direction of airflow, the deposition of fibrous particles in the respiratory tract is largely a function of fibre diameter, with length and aspect ratio being of secondary importance. In addition, the shape of the fibres as well as their electrostatic charge may have an effect on deposition (Davis et al., 1988).

Since most of the data on deposition have been obtained in studies on rodents, it is important to consider comparative differences between rats and humans in this respect; these differences are best evaluated on the basis of the aerodynamic diameter. The ratio of fibre diameter to aerodynamic diameter is approximately 1:3. Thus, a fibre measured microscopically to have a diameter of 1 μm would have a corresponding aerodynamic diameter of approximately 3 μm. A comparative review of the regional deposition of particles in humans and rodents (rats and hamsters) has been presented by US EPA (1988). The relative distribution between the tracheobronchial, and pulmonary regions of the lung in rodents follows a pattern similar to human regional deposition during nose breathing for insoluble particles with a mass median aerodynamic diameter of less than 3 μm. Figures 2 and 3 illustrate these comparative differences. As can be seen, particularly for pulmonary deposition of particles, the percentage deposition in the rodent is considerably less, even within the overlapping region of respiratory tract deposition, than in humans. These data indicate that, although particles with an aerodynamic diameter of 5 μm or more may have significant deposition efficiencies in man, the same particles will have extremely small deposition efficiencies in the rodent.

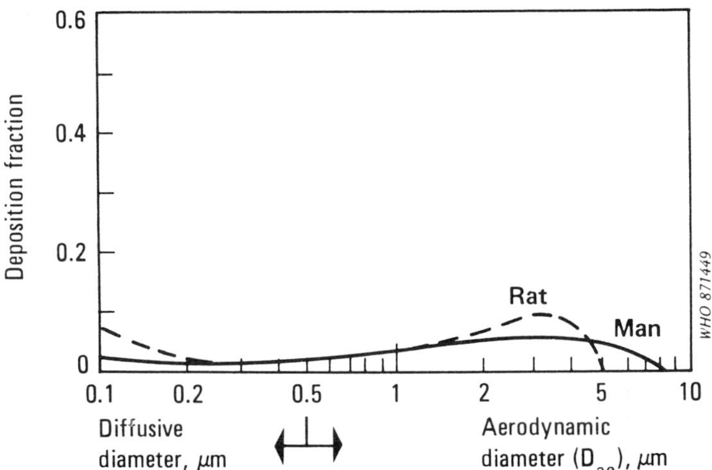

Fig. 2. Tracheobronchial deposition of inhaled monodisperse aerosols in man and rat (US EPA, 1980)

Fibres of various shapes are more likely than spherical particles to be deposited by interception, mainly at bifurcations. Available data also indicate that pulmonary penetration of curly chrysotile fibres is less than that for straight amphibole fibres.

In the nasopharyngeal and tracheobronchial regions, fibres are generally cleared fairly rapidly via mucociliary clearance, whereas fibres deposited in the alveolar space appear to be cleared more slowly, primarily by phagocytosis, to a lesser extent via

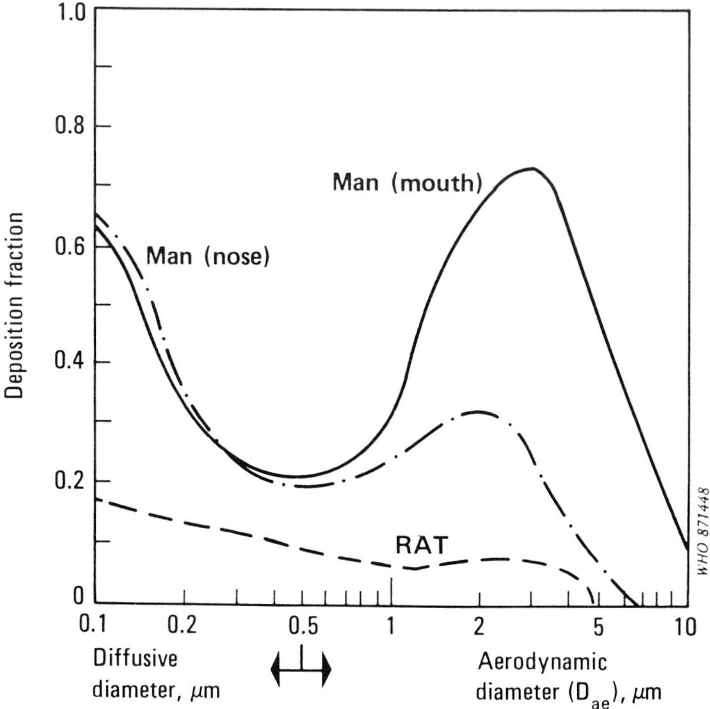

Fig. 3. Pulmonary deposition of inhaled monodisperse aerosols in man and rat (US EPA, 1980)

translocation, and possibly by dissolution. Translocation refers to the movement of the intact fibre after initial deposition at foci in the alveolar ducts and on the ciliated epithelium at the terminal bronchioles. These fibres may be translocated via ciliated mucous movement up the bronchial tree and removed from the lung, or may be moved through the epithelium with subsequent migration to interstitial storage sites or along lymphatic drainage pathways or transport to pleural regions. Fibres short enough to be fully ingested are thought to be removed mainly through phagocytosis by macrophages, whereas longer fibres may be partially cleared at a slower rate either by translocation to interstitial sites, breakage or by dissolution. A higher proportion of longer fibres is, therefore, retained in the lung.

5.2 Studies in experimental animals

Fibres administered in studies on animals are only a subset of those normally present in the occupational environment. Due to the above-mentioned limitations in the anatomy of the rat as a model for man, wherever possible the fraction of rat-respirable fibres in the generated aerosol is specified.

5.2.1 Carbon/graphite fibres

It should be noted that in some of these studies, animals were exposed to fibres that were not respirable for the rodent; in others, animals were exposed principally to dusts, the fibrous content of which was unspecified. For carbon fibres, only one of the studies described here involved exposure to fibres that were respirable for the rat (Warheit et al., in press/a).

In a study designed to investigate the histopathological effects of carbon fibres on the lungs of guinea-pigs, production of a particulate aerosol from a chopped PAN-based carbon fibre ("RAE type 2") proved difficult, although with similar apparatus concentrations of 6000 respirable fibres/ml of asbestos (type unspecified) had been maintained for long periods (Holt & Horne, 1978). The maximum concentration of carbon particles "smaller than 5 μm" produced by the apparatus was 370/ml, with 99% being nonfibrous. Of these, 2.9 fibres/ml ("black fibres") were of respirable size in the dust cloud. These were mainly less than 10 μm in length and about 1 μm in diameter. A total of 15 specific-pathogen-free (SPF) guinea-pigs were exposed to the aerosol described above. At various time intervals up to 104 h post-exposure, groups of animals were sacrificed. Histopathological examination of the lungs of exposed animals revealed macrophages filled with nonfibrous carbon particles, together with a few carbon fibres which were generally extracellular. There were no ferruginous bodies with a carbon core and no pathological effects (Holt & Horne, 1978).

In a follow-up study in which the test atmosphere was similar to that in the investigation described above, groups of SPF guinea-pigs (2-9 per group) inhaled carbon dust for 100 h and were killed at various intervals up to 2 years post-exposure. The respirable fraction of the inhaled dust was mainly nonfibrous and there were very few fibrous particles in sections of the lung. There was some indication, however, that extracellular submicron particles present in the tissue were washed out during histological processing. An occasional ferruginous body was observed. Submicron carbon dust reached the alveoli; phagocytosis of particles began immediately but proceeded slowly over many months and dust-filled macrophages were still evident after 2 years (Holt, 1982).

As described in section 6.1.2.1, a study was conducted in which male Crl:CDBR rats were exposed (nose only) to aerosols of pitch- or PAN-based rat-respirable carbon fibres at target concentrations of 50 or 100 mg/m^3 (47 and 62 fibrils/ml) for periods ranging from 1 to 5 days (6 h/day) and evaluated at 0, 24 and 72 h, 10 days, and 1 and 3 months post-exposure. Pigment-laden alveolar macrophages were observed primarily at the junctions of the terminal bronchioles (Warheit et al., in press/a).

5.2.2 Aramid fibres

Smaller fibrils can peel from the surface of para-aramid fibres as ribbons and complex branching. Because these fine fibrils have unusual shapes and tend to be statically charged, they frequently agglomerate. It is necessary to use specially prepared samples, enriched in respirable fibrils with high-pressure air mills, to generate significant airborne para-aramid fibril concentrations (i.e. > 100 fibres/ml) in inhalation studies on experimental animals.

In a short-term (2 week) inhalation study on rats (as outlined in section 6.1.3.1), Lee et al. (1983) reported that, after 6 months, inhaled Kevlar fibrils accumulated mainly at the bifurcations of the alveolar ducts and adjoining alveoli, with only a few fibrils being deposited in peripheral alveoli of the acinus.

The pattern of deposition and persistence of aramid fibrils in a two-year study was similar to that stated above (Lee et al., 1988). Kevlar fibrils, which are much more curled than chrysotile fibres, were retained mostly in the respiratory bronchioles and alveolar duct region, especially in the ridges of alveolar duct bifurcations. One year after termination of a 12-month exposure to 400 fibrils/ml, the lengths of fibres in lung tissue appeared to be reduced. At the three highest dose levels in the study (25, 100 and 400 fibrils/ml), there was a minute amount of dust accumulation in the alveolar macrophages and in tracheobronchial lymph nodes resulting from the transmigration of intrapulmonary Kevlar fibrils. Most particles in the alveolar macrophages were less than 1 μm long.

In a study by Warheit et al. (1992), Crl:CD rats were exposed (nose only) to ultrafine Kevlar fibrils (6 h/day) for 3 or 5 days at concentrations ranging from 600 to 1300 fibrils/ml (gravimetric concentrations ranging from 2 to 13 mg/m^3) and evaluated at 0, 24, 72 and 96 h, 1 week, and 1, 3 and/or 6 months. Kevlar fibres were found to be deposited at alveolar duct bifurcations located nearest the bronchiolar-alveolar junctions. The median lengths and diameters of ultrafine Kevlar samples in the air and in the lungs were virtually identical immediately following exposure. There was no morphological evidence that fibrils had translocated to epithelial or interstitial compartments, in contrast to patterns observed with chrysotile asbestos. Fibre clearance studies demonstrated a transient increase in the numbers of retained fibrils at 1 week post-exposure, with rapid clearance of fibres thereafter. The transient increase in the number of fibres could have been due to transverse cleaving of the fibres, since the average lengths of retained fibres continued to decrease over time. In this respect, a progressive decrease in the mean length (12.5 to 7.5 μm) and diameter (0.33 to 0.24 μm) of inhaled fibres was measured over a 6-month post-exposure period. The percentage of fibres longer than 15 μm decreased from 30% at time 0 to < 5% at 6 months post-exposure (Warheit et al., in press/b).

Following intratracheal instillation in rats of 25 mg "Kevlar polymer dust" containing a low but undetermined proportion of fibres considered to be in the respirable range (< 1.5 μm in diameter and between 5 and 60 μm in length) in physiological saline for 21 months, large particles were found in the terminal bronchioles, and smaller particles with dimensions of < 5 μm were present in alveolar ducts (Reinhardt, 1980).

In a study in which 5 mg of Kevlar fibres (fibre size distribution and sample preparation methods unspecified) was injected intraperitoneally into Wistar rats, it was reported that fragments were transported through lymphatic pathways and stored in lymph nodes where they caused inflammatory reactions (Brinkman & Müller, 1989).

5.2.3 Polyolefins

Groups of 22 male Fischer-344 rats were exposed nose-only (6 h/day, 5 days/week) for 90 days to filtered air or to 15, 30 or 60 mg/m^3 polypropylene fibre (99.9% purity; 12.1, 20.1 and 45.8 fibres/ml, respectively, size-selected to have an average diameter of 1.6 μm (46% < 1 μm) and an average length of 20.4 μm). There was a strong association between the administered concentration, the time of exposure and the lung fibre burden. Although the length and diameter did not change during the study, the authors hypothesized that the segmentation of these fibres on SEM stubs may have resulted from chemical alteration or partial dissolution of sections of the fibres within the lungs, which made these sections dissolve further during final processing for SEM analysis. This segmentation increased with the administered concentration and period of exposure, as well as with the period of recovery after termination of exposure at 90 days (Hesterberg et al.,1992).

5.3 *In vitro* solubility studies

In an investigation in which the solubility of various natural and synthetic fibres in physiological Gamble's solution (at 37 °C or more for 1 h to 20 weeks and 1 h to 2 weeks for closed and open system conditions, respectively; pH not specified) was determined by atomic absorption spectrometry, carbon and aramid fibres (source and fibre-size distribution unspecified) were found to be "practically insoluble". There was also no evidence of alteration of the surface during examination by SEM with energy dispersive spectrometry (Larsen, 1989).

In a study by Law et al. (1990), test materials including three polymeric organic fibre compositions (polypropylene, polyethylene and polycarbonate) were compared for solubility in physiological Gamble's solution (pH 7.6). The test materials were subjected to leaching for 180 days in a system that provided a continuous flow through sample holders containing the test fibres. After this period, the fibres were examined by electron microscopy for changes in surface area, total specimen weight and surface characteristics. There was virtually no dissolution and no significant change in surface area. There were only slight weight gains, ranging from 0.08 to 0.5%, and no visible surface changes, in contrast to results obtained for several man-made mineral fibres (MMMFs).

6. EFFECTS ON EXPERIMENTAL ANIMALS AND *IN VITRO* TEST SYSTEMS

6.1 Experimental animals

6.1.1 Introduction

There are several factors that should be considered when evaluating experimental data on the biological effects of synthetic organic fibres. (Several of these factors were discussed in relation to man-made mineral fibres by WHO, 1988). Most importantly, synthetic organic fibres should not be considered as a single entity, except in a very general way. There are substantial differences in the physical and chemical properties (e.g., fibre lengths and chemical composition) of the fibres, and it is expected that these would be reflected in their biological responses. Finally, the fibres used for specific research protocols may be altered to determine the biological effects in experimental animals. In this case, they may not represent the hazard potential in humans exposed to the commercial or degradation products during the manufacture, processing, use and disposal.

Notwithstanding the above comments, there are certain characteristics of synthetic organic fibres that are important determinants of effects on biological systems. The most important of these appear to be fibre size (length, diameter, aspect ratio, shape), biopersistence[a] and durability[a], chemical composition, surface area and chemistry, electrostatic charge and number or mass of fibres (dose).

Other considerations relate to the extrapolation of experimental findings for hazard assessment in man. It appears that if a given

---

[a] The term *biopersistence* refers to the ability of a fibre to stay in the biological environment where it was introduced. The term is of particular use in inhalation and intratracheal instillation studies, because a large percentage of the fibres that reach the lung are removed by pulmonary clearance and relatively few are retained (*persist*). The length of time that fibres persist in the tissue is also a function of their *durability*, which is directly related to their chemical composition and physical characteristics. The term *solubility*, as used here, relates to the behaviour of fibres in various fluids. In general, the term *solubility* is more appropriate for use in *in vitro* than *in vivo* studies, because the degradation of fibres in tissues is not only a function of their solubility. While the concept of biopersistence is important, quantitative procedures for evaluation of this parameter have not been established.

fibre comes into contact with a given tissue in animals producing a response, a similar biological response (qualitative) might be expected in humans under the same conditions of exposure. There is no evidence that the biological reaction to fibres differs between experimental animals and humans. However, there may be quantitative differences between species, some being more sensitive than others.

There has been a great deal of debate concerning the relevance of various routes of exposure in experimental animal studies to hazard assessment in man (McClellan et al., 1992). The advantages and disadvantages of each of these routes are discussed in the following sections. It was the consensus of the Task Group that the results of studies by all routes of administration should be considered in evaluating the weight of evidence in hazard evaluation.

Each route cannot be discussed in detail here, but some general observations can be made. Positive results in an inhalation study on animals have important implications for hazard assessment in man. Strong scientifically based arguments would need to be made against the relevance of such a finding to man. Conversely, the lack of a response in an inhalation study does not necessarily mean that the material is not hazardous for humans. Rats, being obligate nose-breathers, have a greater filtering capacity than humans. However, if it were demonstrated that the "target tissue" was adequately exposed and that a biologically important response was not noted, then such a result would be of value for hazard assessment in humans.

As discussed in the Environmental Health Criteria for Man-made mineral fibres (WHO, 1988), a negative result in studies using non-physiological exposure conditions (e.g., intratracheal instillation, intrapleural injection or implantation, intraperitoneal injection) would suggest that a specific fibre may not be hazardous for parenchymal lung tissue and/or the mesothelium. In contrast, a positive result in such studies should be confirmed by further investigation in inhalation studies for a complete assessment of the hazard for humans.

However, the Task Group believes that the use of instillation/ injection studies may not necessarily be appropriate for certain synthetic organic fibres. This recommendation is based on the lack of, or weak response to, para-aramid fibres seen in three intraperitoneal studies (Pott et al., 1987, 1989; Davis, 1987), whereas a definite neoplastic response was seen in a chronic inhalation study (Lee et al., 1988) (see also Appendix 1). The authors of these intraperitoneal studies speculated that the lack of response to para-aramid fibres may be due to the ability of this type of fibre to agglomerate and thereby reduce the actual number of single fibres. These results suggested to the Task Group that a negative result in an instillation/injection study with selected synthetic organic fibres should not always be considered to indicate an absence of hazard in humans by inhalation exposure.

In vivo dissolution and translocation studies play an important part in our understanding of the behaviour of fibres in the lung. Such studies (Bernstein et al., in press) of comparative biopersistence can play a part in hazard identification.

In several studies, the effects of prostheses composed of various synthetic organic fibres on tissues have been examined (Neugebauer et al., 1981; Tayton et al., 1982; Makisato et al., 1984; Parsons et al., 1985; Henderson et al., 1987). These studies have not been reviewed here since they are not relevant to an assessment of the effects of inhaled synthetic organic fibres.

6.1.2 Carbon/graphite fibres

There are no adequate studies available in which the fibrogenic or carcinogenic potential of carbon/graphite fibres have been examined.

6.1.2.1 Inhalation

In a subchronic inhalation study, a group of 60 male Sprague-Dawley rats was exposed to pulverized carbon fibres (20 mg/m^3; bulk product PAN-based Celion fibres with mean diameter of 7 μm and 20-60 μm in length; aerosol not characterized), 6 h/day, 5 days/week for up to 60 weeks, and a similarly sized control group was exposed to air alone (Owen et al., 1986). There was a slight decrease in the rate of body weight gain in the first 4 weeks of exposure. In the post-exposure period, however, the average body weight of the exposed rats was slightly but not significantly less than that of the control animals. Although there were variable changes in the average airway resistance for inspiration, and the respiratory rate and minute volume of the exposed animals were significantly less than those of the controls, these changes were not considered by the authors to be related to exposure to carbon fibres.

In animals exposed to carbon fibres, there was a low-grade, diffuse increase in alveolar macrophages containing fibrous particles in the lungs but no pulmonary fibrosis or inflammatory reaction. By 32 weeks post-exposure, there were only occasional alveolar macrophages containing fibres or particles scattered throughout the lungs (Owen et al., 1986). The Task Group noted that the lack of observed effects in the lungs may be attributable to the fact that the administered fibres were not in the respirable size range for the rat.

Groups of unspecified numbers of male Crl:CDBR rats were exposed (nose only) to aerosols of pitch- or PAN-based respirable carbon fibres at target concentrations of 50 or 100 mg/m^3 for periods ranging from 1 to 5 days (6 h/day) and evaluated at 0, 24 and 72 h, 10 days, and 1 and 3 months (Warheit et al., in press/a). A 5-day exposure to respirable pitch-based carbon fibres (47 or 106 mg/m^3; 47 and 62 fibres/ml, probably $\geq$ 5 μm in length; MMADs of 1.3 μm and 1.6 μm, respectively) produced dose-dependent, transient inflammatory responses in the lungs of exposed rats, manifested by increased levels of neutrophils and concomitant significant increases in lactate dehydrogenase, protein or alkaline phosphatase in bronchoalveolar fluids at early post-exposure time periods. These changes were reversible within 10 days after exposure. There were no significant differences in the morphology or *in vitro* phagocytic capacities of macrophages recovered from rats between the sham-exposed control group and those exposed to pitch-based carbon fibres. Results from cell labelling studies in rats exposed to pitch-based carbon fibres for 5 days demonstrated an increased turnover of lung parenchymal cells at 10 days or 1 month after exposure, which did not correlate with the measures of inflammation in bronchoalveolar fluids. No increases in turnover of terminal bronchiolar cells were measured at any time post-exposure. Pigment-laden alveolar macrophages and minimal type II epithelial cell hyperplasia were observed primarily at the junctions of the terminal bronchioles and alveolar ducts. In an additional group of rats used as negative controls, exposed for 6 h to PAN-based carbon fibres (diameter outside the respirable range; MMAD $\geq$ 4.4 μm), there were no cellular, cytotoxic or alveolar/capillary membrane permeability changes at any time post-exposure.

6.1.2.2 Intratracheal administration

Following characterization of dusts from machined composites containing five graphite fibres and one fibrous glass, as described in section 4.1.1.2 (Boatman et al., 1988), groups of five male pathogen-free Charles River rats received a single intratracheal injection of 5 mg of human respirable fractions of the dusts in sterile, pyrogen-free phosphate-buffered saline (Luchtel et al., 1989). The graphite composites included in the study were a proprietary material designated as graphite-p, two graphite materials manufactured from PAN, one pitch-based graphite composite and graphite-PAN/Kevlar. The machining operations included spindle shaping (3450 or 10 000 rpm), hand routing (23 000 rpm) and use of a saber saw. The mean particle diameter of tested fractions was 2.7 μm, with 74% being less than 3.0 μm in diameter (99.9% < 10 μm in aerodynamic diameter). For comparison, similarly sized groups of animals were exposed to phosphate-buffered saline, aluminium oxide (negative control) and quartz (positive control). No information on the fibre content of the dust samples was presented.

In a study by Martin et al. (1989), one of the lungs of each rat was examined histopathologically one month following injection and the other was lavaged to recover airway cells and fluid. For the six composite-epoxy materials, there was a continuum of lung response (from an increase in alveolar macrophages to fibrosis) that fell between that observed in the positive (quartz) and negative (non-fibrous aluminium oxide) control groups. None of the composite dusts induced effects that were as severe as those observed with quartz. However, four of the dusts (fibrous glass composite, the two graphite materials manufactured from PAN and the pitch-based graphite composite) produced responses (discussed below) that were more severe than those to aluminium oxide. Among the composite materials, the reactions for one of the PAN-based graphite and fibrous glass composites were the most severe. The pitch-based graphite and one of the PAN-based composites caused the greatest increase in total cells in lung lavage fluid. However, total cells recovered in animals treated with any of the composite samples did not differ significantly from those for animals treated with either NaCl or aluminium oxide. In contrast, the quartz-treated animals had ten times more lavaged cells than the NaCl-treated animals and five times more than any of the animals treated with composite samples. There were also highly significant differences in the types of cells present in the lavage fluid among the various groups, with the same pitch-based graphite and one of the PAN-based composites producing the greatest increase in both percentage and total number of lavaged neutrophils among the composites tested. The percentage and the absolute number of neutrophils were greater in the quartz-treated animals than in any other group.

The authors noted that bolus administration by intratracheal instillation may overwhelm lung defence mechanisms, but felt that these results raised the possibility that some types of composite dusts may be fibrogenic in humans (Luchtel et al., 1989).

6.1.2.3 Intraperitoneal administration

Groups of twelve albino Wistar rats (six male and six female) were injected intraperitoneally with "carbon dust" (50 mg/kg body weight) suspended in physiological saline at a concentration of 10 mg/ml. Particle diameters determined by electron microscopy ranged from 0.2 to 15 μm. The particle concentration in the injected solution measured by haemocytometer (resolution power, 1 μm) was 3.75 x 10^6 (reported as x 10^{-6}) per ml suspension containing 0.5 mg of dust. At 1 and 3 months after administration, there were no treatment-related lesions in tissues examined histopathologically (i.e. omentum, spleen, liver and pancreas), whereas in U.I.C.C. Rhodesian chrysotile-exposed animals (12.5 mg/kg body weight at 2.5 mg/ml in physiological saline) there were characteristic fibrotic nodules in the peritoneum (Styles & Wilson, 1973). The Task Group noted that in the published account of this study it was not clear whether the material was fibrous, particulate or both.

6.1.2.4 Dermal administration

In a study reported by Depass (1982), four types of carbon fibres (continuous pitch-based filament, pitch-based carbon fibre mat, polyacrylonitrile continuous fibres and oxidized PAN-based fibres) were ground and suspended in benzene (25 μl of a 10% (w/v) suspension) and applied to the clipped skin of the back of 40 male C3H/HeJ mice three times weekly until death. No statistically significant increases in skin tumours were observed in any of the exposed groups compared to controls receiving vehicle alone. The Task Group considered this study inadequate for an evaluation due to the lack of reporting data regarding the nature of the materials including the particle size and morphology of the test materials.

6.1.3 Aramid fibres

6.1.3.1 Inhalation

A summary of the design of the inhalation studies with aramid fibrils is presented in Table 6.

Table 6. Inhalation studies on aramid fibres

Fibre type	Concentrations of fibrils		Size distribution[c,d]	Exposure type and duration	Histopathological evaluation (sacrifices)	Number and strain of animal	Reference
	Mass (mg/m^3)	Number (per ml)					
Para-aramid dust with 2% elongated (fibres)	150	low proportion < 1.5 μm		whole body; acute, 4 h	not specified (14 day)	not specified (rat)	Reinhardt (1980)
	130	not specified	not specified	4 h/day, 5 days/ week, 2 weeks	0 and 14 days post-exposure	not specified (rat)	
Kevlar fibrils	0	0	(length: 53% > 20 μm)	whole body; 6 h/day, 5 days/ week, 2 weeks	0, 2 weeks 3 months 6 months (post-exposure)	5 male rats/group Crl:CD Sprague-Dawley (Charles River)	Lee et al. (1983)
	0.1 ± 0.06	1.3	85% < 5 μm MMAD				
	0.52 ± 0.14	26	87% < 5 μm MMAD				
	3.0 ± 0.4	280	94% < 5 μm MMAD				
	18.2 ± 2.8	not determined	49% < 5 μm MMAD				
	17.6[a] ± 4.4	not determined	13% < 5 μm MMAD				

Table 6 (contd).

Fibre type	Concentrations of fibrils		Size distribution[c,d]	Exposure type and duration	Histopathological evaluation (sacrifices)	Number and strain of animal	Reference
	Mass (mg/m^3)	Number (per ml)					
Kevlar fibrils	0 0.08 ± 0.04 0.32 ± 0.08 0.63 ± 0.14 2.23 ± 0.46	0 2.4 ± 0.80 25.5 ± 9.9 100 ± 37 411[b] ± 109	more than 70% by mass, < 5 μm MMAD	whole body; 6 h/day, 5 days/ week, 2 years	3 months 6 months 12 months; final sacrifice at 24 months (no recovery period after 24 months)	100 male and 100 female per group Crl:CD(SD)BR (Charles River)	Lee et al. (1988)
Kevlar fibrils	2.0-13.3	600-1300	3.2-4.7 μm MMAD CMC = 10 μm CMD = 0.3 μm	nose only; 3 or 5 days	0 24 h 72 h 96 h 1 week 1, 3, 6 months (post-exposure)	4 males per group Crl:CDBR (Charles River)	Warheit et al. (1992)

a   Commercial Kevlar sample
b   Exposure was for 1 year only due to toxic response. Surviving animals were held for an additional 1 year without exposure.
c   It should be noted that the rat can inhale into the distal airways fibres with diameters less than 1 to 1.5 μm.
d   MMAD = mass median aerodynamic diameter

Reinhardt (1980) reported a summary of the results of an acute and a 2-week inhalation study in rats exposed to a mixture of paraaramid dust containing 2% elongated particles with an aspect ratio of at least 3:1 (fibre). Details provided in the published account of these studies were insufficient for evaluation. In the short-term study, rats (number and strain unspecified) were exposed to Kevlar dust (130 mg/m^3) 4 h/day, 5 days/week, for 2 weeks. Control rats were exposed to filtered air concurrently. Following the final exposure, half of both the test and control group animals were sacrificed and 21 tissues examined histopathologically. The remaining rats were sacrificed and examined similarly following a 14-day recovery period. During exposure, test animals were slightly less active and gained less weight than controls, effects which were not observed during the recovery period. In rats examined after the tenth exposure, there were numerous macrophages in the lung tissue. At the end of the recovery period, these macrophages decreased in number and formed discrete clusters, which the author interpreted as a "non-specific response to foreign particles in the lung" (Reinhardt, 1980). The Task Group noted that available information presented in the published account of this study was insufficient to determine the fraction of the aerosol which was respirable for the rat.

In a short-term inhalation study, groups of male Sprague-Dawley rats (number in each group unspecified) were exposed to 0, 0.1 mg/m^3 (1.3 fibrils/ml), 0.5 mg/m^3 (26 fibrils/ml), 3.0 mg/m^3 (280 fibrils/ml) or 18 mg/m^3 (number concentration not determined) of ultrafine Kevlar pulp fibrils prepared specially by a high-pressure air impingement device (60-70% < 1 μm in diameter and between 10 and 30 μm long) for 6 h/day, 5 days/week, for 2 weeks. Another group was exposed to 18 mg/m^3 commercial Kevlar fibres (2.5 mg/m^3 of respirable dust) for the same period. Five rats in each group were killed at the end of exposure and at several periods up to 6 months after exposure. In animals exposed to 0.1 or 0.5 mg/m^3 of the ultrafine fibres there was a macrophage response in the alveolar ducts and adjoining alveoli, which was almost completely reversible within 6 months after exposure. In rats exposed to 3 mg/m^3 ultrafine fibres or 18 mg/m^3 of the commercial product, there was occasional patchy thickening of alveolar ducts with dust and inflammatory cells (but no collagen) 6 months after exposure. In the group exposed to 18 mg/m^3 ultrafine Kevlar, there were granulomatous lesions with dust cells in the respiratory bronchioles, alveolar ducts and adjoining alveoli after two weeks of exposure. One month following exposure, there was patchy fibrotic thickening in the alveolar duct regions and adjacent alveoli, as well as dust cells. The fibrotic lesions were markedly reduced in cellularity, size and numbers from 3 to 6 months after exposure but contained networks of reticulum fibres with some collagen fibres (Lee et al., 1983). (It should be noted that there is some discrepancy between the results presented in the published account of this study and a summary of this study included in the report of the longer-term bioassay (Lee et al., 1988 described below).

Groups of 24 male Crl:CDBR rats were exposed (nose only) to ultrafine Kevlar fibres (fibrils) 6 h/day for 3 or 5 days at concentrations ranging from 600 to 1300 fibres/ml (gravimetric concentrations ranging from 2 to 13 mg/m^3) and subgroups of four rats were subsequently evaluated at 0, 24, 72 and 96 h, 1 week, and 1, 3 and/or 6 months post-exposure (Warheit et al., 1992; in press/b). The authors suggested that at higher gravimetric concentrations, there was probably an agglomeration of fibres in the aerosols. Five-day exposures elicited a transient granulocytic inflammatory response with an influx of neutrophils into alveolar regions and concomitant increases in bronchio-alveolar lavage fluid levels of alkaline phosphatase, lactate dehydrogenase and protein, which returned to control levels at time intervals of between 1 week and 1 month post-exposure. Macrophage function (as determined by surface morphology and *in vitro* phagocytic and chemotactic capacities) in Kevlar-exposed alveolar macrophages was not significantly different from that of sham controls at any time interval. Increased pulmonary cell labelling was noted in terminal bronchiolar cells immediately after exposure but values returned to control levels one week later. Histopathological examination of the lungs of Kevlar-exposed animals revealed only minor effects, characterized by the presence of fibre-containing alveolar macrophages situated primarily at the junctions of terminal bronchioles and alveolar ducts.

In the only chronic inhalation study conducted for any of the organic fibres considered in this document, groups of 100 male and female Crl:CD(SD)BR weanling rats were exposed to ultrafine Kevlar fibrils at concentrations of 0, 2.4, 25.5, and 100 fibrils/ml (0, 0.08, 0.31 and 0.63 mg/m^3, respectively), 5 days/week for two years (Lee et al., 1988). An additional group of 100 animals was exposed according to the same schedule to 400 fibrils/ml (2.23 mg/m^3), and, owing to toxicity, exposure was terminated at 12 months and the animals were followed for an additional year. There were interim sacrifices of 10 males and 10 females per group at 3, 6 or 12 months. Fibrils were separated from the Kevlar pulp matrix by high-pressure air impingement as described above for the study conducted by Lee et al. (1983). A summary of the effects on the lung is presented in Table 7. At a concentration of 2.5 fibrils/ml, the alveolar architecture of the lungs was normal with a few dust-laden macrophages in the alveolar airspaces (alveolar macrophage response was considered by the authors to be the no-observed-adverse-effect level). At 25 and 100 fibrils/ml, there was a dose-related increase in lung weight, a dust cell response, slight type II pneumocyte hyperplasia, alveolar bronchiolarization and a "negligible" amount of collagenized fibrosis in the alveolar duct region. In addition, at 100 fibrils/ml, "cystic keratinizing squamous cell carcinomas" (CKSCC; tumours not observed spontaneously in this strain or in man) were found in four female rats (6%) but not in any male animals. Female rats also had more prominent foamy alveolar macrophages, cholesterol granulomas and alveolar bronchiolar-ization, and this was related to the development of CKSCC. At 400 fibrils/ml, 29 male and 14 female rats died due to obliterative bronchiolitis resulting from dense accumulation of inhaled Kevlar fibrils in the ridges of alveolar duct bifurcations during exposure for one year to 400 fibrils/ml. In the case of animals surviving one year post-exposure, the lung dust content, average fibre length and the pulmonary lesions in surviving rats were markedly reduced, but there were slight centriacinar emphysema and minimal fibrosis in the alveolar duct region. One male (1/36; 3%) and six female (6/56; 11%) rats in this experimental group developed CKSCCs.

The CKSCCs developed between 18 and 24 months of age. The investigators reported that microscopically, it was extremely difficult to distinguish between squamous metaplasia and CKSCC since the lung tumours were well differentiated and there was no evidence of either tumour metastasis or invasion to adjacent tissue. The tumours were, therefore, considered by the authors to be benign neoplastic lesions which were classified as CKSCC due to the fact that there is no benign type of squamous cell lung tumour widely accepted in humans. The investigators also suggested that these changes should perhaps be considered as either metaplastic or dysplastic rather than as neoplastic lesions. In instillation studies in rats with silica, however, CKSCCs have been described as precursors of squamous cell carcinomas, which develop if the life span is sufficiently long (Pott et al., in press). The Task Group members noted that there is considerable debate concerning the biological potential of these lesions (CKSCC) and their

Table 7. Summary of effects on the lungs of chronic inhalation of aramid fibrils[a]

Exposure concentration (fibrils/ml)	Mortality	Lung fibrosis	Keratinized squamous cell metaplasia	Cystic keratinizing squamous cell carcinoma (CKSCC)[b]	Adenomas
0	0	-	-	-	1/69 (1%) male 0/68 (0%) female
2.5	0	-	-	-	1/69 (1%) male 0/64 (0%) female
25	0	dose-related increase in severity and incidence	-	-	1/67 (2%) male 0/65 (0%) female
100	0		-	0/68 male 4/69 (6%) female	1/68 (2%) male 3/69 (4%) female
400	29/65 male 14/70 female		0/36 male 6/69 (9%) female	1/36 (3%) male 6/56 (11%) female	2/36 (6%) male 2/56 (4%) female

[a] From: Lee et al. (1988)
[b] See also Appendix 1

relevance to humans[a]. It was the view of the Task Group that the CKSCCs were related to exposure to aramid and that these lesions are part of the neoplastic spectrum. However, the Task Group also felt that the high dose (400 fibres/ml) exceeded the maximum tolerated dose. Finally, the Task Group noted that the study was terminated at 24 months; a longer observation time might have yielded a higher incidence of these tumours.

6.1.3.2 Intratracheal administration

In the same brief account referred to in section 6.1.3.1, Reinhardt (1980) reported the protocol and results of a study in which 2.5 mg of shredded Nomex paper in physiological saline was instilled into the trachea of rats (number, strain, instillation schedule and nature of control group not specified). Fibre sizes were reported to vary from 2 to 100 μm in length and 2-30 μm in diameter. The lungs of groups of rats were examined histologically at 2 and 7 days, 3 and 6 months, and at 1 and 2 years. There were no adverse effects other than initial transitory acute inflammation followed by foreign body granuloma formation. These mild tissue reactions became less obvious post-exposure and the lungs were essentially normal without formation of collagenized fibrosis two years after exposure.

In a similar briefly documented study, 25 mg Kevlar in physiological saline was instilled intratracheally into rats (number and strain not specified) and control rats were administered saline alone. Animals were sacrificed 2, 7, and 49 days and 3, 6, 12 and 21 months after treatment and the respiratory tract was examined histopathologically. Following instillation, particles could be detected in lung tissue. An initial, non-specific inflammatory response subsided within about a week and foreign body granulomas with minimal collagen were seen at later sacrifices. Tissue responses decreased with increasing time post-exposure (Reinhardt, 1980).

6.1.3.3 Intraperitoneal administration

In studies by Pott et al. (1987), 5-week-old female Wistar rats were administered 10 mg Kevlar fibres prepared by ultrasonic

---

[a]   After the meeting of the IPCS Task Group an international panel of 13 pathologists evaluated these cystic lesions. The summary of their evaluation and the names of the participants are in Appendix 1.

treatment in three weekly intraperitoneal injections of 2, 4 and 4 mg. At the end of the study (surviving animals sacrificed 2.5 years after treatment), 4 out of 31 animals (12.9%) had tumours (sarcoma, mesothelioma or carcinoma of the abdominal cavity). In an additional study in which there was an attempt to obtain finer fibres and better suspension by drying, milling and ultrasonic treatment, 20 mg Kevlar (50% < 3.4 μm in length and 50% < 0.47 μm in width) in saline was injected intraperitoneally into 8-week-old Wistar rats (5 injections of 4 mg weekly). At 28 months after injection, the percentage of tumour-bearing animals was 5.8 (preliminary results; 34 animals sacrificed and 18 survivors). The authors commented that it was difficult to produce a homogeneous suspension of Kevlar fibres and that, as a result, these fibres were more likely to be present in clumps in the peritoneal cavity than were other dusts. In these studies, tumour incidences after injection of 0.25 to 0.5 mg actinolite, chrysotile, crocidolite or erionite were between 50 and 80%, and, of 204 rats injected intraperitoneally with saline alone, 5 (2.5%) had malignant tumours in the abdominal cavity. The Task Group noted that there were some discrepancies between the reported results of this investigation and the later study by Pott et al. (1989). Moreover, it was unclear whether these references refer to the same or different studies.

In a subsequent report by Pott et al. (1989), in which both the fibre size distribution (90% of the fibre diameters in the administered material were < 0.76 μm, and 50% of fibre lengths were > 4.9 μm and 90% were < 12 μm) and number of fibres were characterized, there was no significant increase in peritoneal tumours. There were tumours in 3 out of 53 (5.7%) female Wistar rats compared to 2 out of 102 (2%) in the controls at 130 weeks following intraperitoneal administration of 4 weekly doses of 5 mg of milled bulk Kevlar (total dose, 20 mg). The number of Kevlar fibres administered was 1260 x 10^6. In contrast, following administration of a much smaller total dose of UICC chrysotile (0.25 mg, 202 x 10^6 fibres), tumour incidence was 68% (Pott et al., 1989).

Based on an examination of two animals from the study of Pott et al. (1989), Brinkman & Müller (1989) described the following stages of events following intraperitoneal injection of 5 mg of Kevlar fibres (fibre size distribution or sample preparation methods not specified) suspended in 1 ml physiological saline injected into 8-week-old Wistar rats at weekly intervals for 4 weeks. At 28 months after the first injection, the rats were sacrificed and the greater omentum with pancreas and adhering lymph nodes were removed and examined histologically by light and scanning electron microscopy. In an initial stage, there were multinucleated giant cells with phagocytosis of the Kevlar fibres and an inflammatory reaction. In a second stage, granulomas with central necrosis developed indicating the cytotoxic nature of the fibres. A third stage was characterized by "mesenchymal activation with capsular structures of collagenous fibres as well as a slight submesothelial fibrosis". Finally, the reactive granulomatous changes in the greater omentum of the rats were accompanied by proliferative mesothelial changes which, in one of the two animals examined, led to mesothelioma. The authors commented that the reaction to Kevlar in the intraperitoneal test resembled the well-studied reaction to similar injections of glass or asbestos fibres. It was also noted that, as in the case of mineral fibres, fragments of Kevlar fibres were transported through lymphatic pathways and stored in lymph nodes where they caused inflammatory reactions.

In a study in which 25 mg of Kevlar (fibre size distribution unspecified) were administered intraperitoneally to Sprague-Dawley rats (20 of each sex), there were no peritoneal mesotheliomas at termination (104 weeks) (Maltoni & Minardi, 1989). In an additional study by the same investigators, there were no peritoneal mesotheliomas at 76 weeks in similarly sized groups of rats of the same strain following intraperitoneal administration of 1, 5 or 10 mg Kevlar fibres (fibre size distribution unspecified) (Maltoni & Minardi, 1989).

Doses of either 0.25, 2.5 or 25 mg of Kevlar pulp vigorously disaggregated by a turreted tissue homogenizer (96% with diameters < 1 μm; 56% with diameters < 0.25 μm) were administered by single intraperitoneal injection in phosphate buffered saline to three groups of 3-month-old male AF/Han strain rats comprising 48, 32 and 32 animals, respectively (Davis, 1987). An additional group of 12 animals was injected with 25 mg of disaggregated Kevlar pulp and killed at intervals of between 1 week and 9 months after injection to examine the early histopathological reaction. A group of 48 untreated rats was maintained as a control. The authors commented that it was not possible to report the fibre length distribution or fibre number concentration since it was often not possible to determine whether disaggregated fibrils were still attached or simply tangled with the larger fibres in the material prepared for injection. Although the number of separate free fibrils greatly exceeded that of the larger fibres present, the latter most probably made up the bulk (by mass) of the injected material. Consequently, the number of fibrils per unit mass injected that were within the size range normally considered to be most potent in the induction of mesothelioma was much lower than for most asbestos or man-made mineral fibre preparations examined to date. (The dust generation technique used in the studies of Lee et al. (1983, 1988) produced a much finer preparation than was used in the current study).

There were no significant differences in survival between the exposed and control groups. The cellular reaction to injected Kevlar was considered to be vigorous with development of large cellular granulomas. Although not a significant increase, 2 out of 32 animals in the highest dose group (25 mg) developed peritoneal mesotheliomas and it was concluded that the Kevlar preparation possessed a low but definite carcinogenic potential (Davis, 1987).

6.1.4 Polyolefin fibres

There are no adequate studies in which the fibrogenic or carcinogenic potential of polyolefin fibres have been examined.

6.1.4.1 Inhalation

In a study by Hesterberg et al. (1992), groups of 22 male Fischer-344 rats were exposed nose-only (6 h/day, 5 days/week for 90 days) to filtered air or to 15, 30 or 60 mg/m^3 polypropylene fibre (99.9% purity; 12.1, 20.1 and 45.8 fibres/ml, respectively, size-selected to have an average diameter of 1.6 μm (46% < 1 μm) and an average length of 20.4 μm). No abnormal clinical signs were observed in any exposure group. There were no statistically significant changes in body or lung weight or excess mortality as compared to the control. Necropsy and histopathological investigations were performed on subgroups of 6 to 10 rats randomly selected from each group immediately after 30 and 90 days of exposure and 30 days after the 90-day exposure was terminated. At all time points in the study there were dose- and duration-dependent changes in the lungs characterized by increased cellularity and early bronchiolitis but no deposition of collagen. These cellular changes appeared to be reversible at the lower dose levels 30 days post-exposure. There was a strong association between the administered concentration, the time of exposure and the lung fibre burden.

6.1.4.2 *Intratracheal administration*

In a study (summary report) by M.B. Research Laboratories (1980), single doses (unspecified) of ozonized (i.e. charge neutralized) "polyethylene SHFF", ozonized "polypropylene SHFF" or "HHF polypropylene" (source and fibre size distribution unspecified) were administered by intratracheal insufflation in Tween 60 to groups of 40 male Long-Evans rats. No effects on the lung were reported, but the Task Group considered that the data presented in this report were insufficient for evaluation.

6.1.4.3 *Intraperitoneal administration*

Groups of 12 Wistar rats (Alderley Park strain; 6 male and 6 female) were injected intraperitoneally with a single dose of 50 mg/kg (5 ml; 10 mg/ml) of either polyethylene ("Alkathene") or polypropylene dusts in physiological saline (Styles & Wilson, 1973). Particle diameters determined by electron microscopy were 3 to 75 μm (polyethylene) and 4 to 50 μm (polypropylene). Particle concentrations in the injected solution measured by haemocytometer (resolution power, 1 μm) in 1 ml suspensions containing 0.5 mg of dust were 2.38 x 10^6 (reported as x 10^{-6}) for polyethylene and 1.94 x 10^6 (also reported as x 10^{-6}) for polypropylene. At 1 and 3 months, animals were sacrificed and the omentum, spleen, liver and pancreas examined histopathologically. No treatment-related lesions were observed, whereas in U.I.C.C. Rhodesian chrysotile-exposed animals (12.5 mg/kg body weight at 2.5 mg/ml in physiological saline) there were characteristic fibrotic nodules (Styles & Wilson, 1973). It was not clear to the Task Group whether this material was fibrous, particulate or both.

No significant increase in peritoneal tumours was observed in one intraperitoneal study, reported by Pott et al. (1987, 1989), where 10 mg of polypropylene fibres (50% < 7.4 μm in length and 50% < 1.1 μm in diameter) in saline was injected intraperitoneally into 8-week-old Wistar rats once a week for 5 weeks (total, 50 mg). At 28 months after injection, 4% (2 out of 51) of the animals had tumours (sarcoma, mesothelioma or carcinoma of the abdominal cavity) (Pott et al., 1989). In other studies, tumour incidences after injection of 0.25 to 0.5 mg actinolite, chrysotile, crocidolite or erionite were between 50 and 80%; 2 out of 102 rats injected intraperitoneally with saline alone had malignant tumours in the abdominal cavity (Pott et al., 1987, 1989). The Task Group noted that there were some discrepancies between the reported results of this investigation and the later study by Pott et al. (1989). Moreover, it was not clear whether these references referred to the same or different studies.

6.2 *In vitro* studies

In vitro short-term studies, e.g., cytotoxicity, cytogenicity, and cell transformation studies, contribute to an understanding of the mechanisms of action of fibres. The results of such studies are useful in the overall assessment of fibre toxicity, but should not be used alone for hazard assessment. However, it should be noted that there are no known negative controls for *in vitro* studies with fibrous materials.

6.2.1 Carbon fibres

Martin et al. (1989) evaluated the *in vitro* effects of a series of five graphite fibre composite materials machined by various operations (as characterized by Boatman et al. (1988) and described in section 4.1.1.2) in rabbit alveolar macrophages by trypan blue exclusion, release of ^{51}Cr from prelabelled macrophages and phagocytosis as measured by light microscopy. The Task Group noted that it was not clear in the report whether this material was fibrous, particulate or both. Approximately 74% and 99.9% of the particles in each sample were less than 3.0 μm and 10 μm in aerodynamic diameter, respectively. For comparison, a fibreglass composite material, aluminium oxide (negative control) and alpha quartz (positive control) were also tested. Following administration of 500 μg/ml, two of the samples ("graphite-PAN with epoxy and aromatic amine curing agent with geometric mean particle diameter of 1.8 μm" and "graphite-pitch with epoxy and aromatic amine curing agent with geometric mean particle diameter of 1.6 μm") were consistently the most cytotoxic producing the greatest release of ^{51}Cr from labelled alveolar macrophages and the greatest reduction in viability based on trypan blue exclusion. The cytotoxicity of "fibreglass with epoxy and amine curing agent with geometric mean particle diameter of 2.6 μm" and "graphite-PAN with epoxy and amine curing agent with geometric mean particle diameter of 1.1 μm" was similar to that of the negative control (aluminium oxide). The cytotoxicity of "graphite-p with nonepoxy polyetherketone thermoplastic with geometric mean particle diameter of 1.6 μm" and "graphite-PAN/Kevlar with epoxy and amine curing agent with geometric mean particle diameter of 1.9 μm" was intermediate.

In studies conducted by Styles & Wilson (1973), "carbon dust" was not considered cytotoxic in rat alveolar and peritoneal macrophages. Less than 2% of peritoneal macrophages and 5% of alveolar macrophages were killed following phagocytosis of carbon dust with particle diameters determined by electron microscopy to range from 2 to 15 μm. Particle concentrations measured by haemocytometer (resolution power, 1 μm) in 1-ml suspensions containing 0.5 mg of dust were 3.75 x 10^6 (reported as x 10^{-6}). Cell cultures were incubated for 2 h with an unspecified volume of a stock suspension containing 150 mg of dust per ml in (BSS); samples were taken at 0, 1, and 2 h after the addition of the dust (Styles & Wilson, 1973).

"Acetone reconstituted benzene extracts" of two carbon fibre types (pitch-based and PAN-based carbon fibres) were tested in a series of genotoxicity assays (US EPA, 1988). The test materials were not mutagenic but were weakly clastogenic. The Task Group considered that these studies are of little or no value for an evaluation since no information was presented concerning the nature of the test substances.

6.2.2 Aramid fibres

Aqueous solutions containing 25, 50, 100 and 250 μg/ml Kevlar, extracted from commercial grade Kevlar by dispersion and settling in distilled water (90% $\leq$ 5 μm in length and 0.25 μm in diameter; average length and diameter, 2.72 and 0.138 μm, respectively), were cytotoxic to pulmonary alveolar macrophages obtained from adult male Long-Evans black hooded rats, based on determination of leakage of cytoplasmic lactic dehydrogenase, lysosomal enzymes, beta-galactosidase, and ATP content (incubation time, 18 h). The cytotoxic response in freshly harvested and cultured cells was considered to be similar to or greater than that for UICC B Canadian chrysotile (Dunnigan et al., 1984). The Task Group noted that due to their short length, these fibres would not be included in fibre counts in the occupational setting, according to WHO criteria (WHO, 1985).

6.2.3 Polyolefin fibres

In studies conducted by Styles & Wilson (1973), polyethylene ("Alkathene") and polypropylene were considered, on the basis of cytotoxicity in rat alveolar and peritoneal macrophages, to be among the least toxic of various dusts. Less than 2% of peritoneal macrophages and 5% of alveolar macrophages were killed following phagocytosis of polyethylene and polypropylene with particle diameters determined by electron microscopy to range from 3 to 75 μm and 4 to 50 μm, respectively. The Task Group noted that it was not clear in the report whether this material was fibrous, particulate or both. Particle concentrations measured by haemocytometer (resolution limit, 1 μm) in 1 ml suspensions containing 0.5 mg of dust were 2.38 x 10^6 (reported as x 10^{-6}) and 1.94 x 10^6 (reported as x 10^{-6}). Cell cultures were incubated for 2 h with an unspecified volume of a stock suspension containing 150 mg of dust per ml in BSS; samples were taken at 0, 1, and 2 h after the addition of the dust (Styles & Wilson, 1973).

Extracts from three different types of polyethylene granules (each with and without additives) and three polyethylene films (high content of additives) were not mutagenic in *Salmonella typhimurium* strains TA98, TA100 and TA1537 (Fevolden & Moller, 1978). Information included in the published account of this study (limited to an abstract) was insufficient for evaluation.

7. EFFECTS ON HUMANS

The data available on the health effects of synthetic organic fibres in humans are extremely limited. Information is currently limited to case reports and small cross-sectional morbidity studies of workers, without standardized methods and appropriate control groups. The negative results of some of these studies are most likely a function of their limited power to detect an effect, since only relatively small groups of workers with relatively low and short exposures have been examined to date. On the other hand, positive results reported from studies with possibly higher exposures are poorly documented and observed effects may be due, in part, to other substances present in the occupational environment.

7.1 Carbon/graphite fibres

In a cross-sectional study of 88 out of 110 workers in a PAN-based continuous filament carbon fibre production facility, there were no adverse respiratory effects, as assessed by radiographic and spirometric examination (for determination of FEV_1 and FVC) and the replies to questionnaires on respiratory symptoms. Total dust concentrations were 0.08 to 0.39 mg/m^3, with 40% being in the respirable range. Only 31 of the workers examined, however, had been employed for more than 5 years in the facility in which carbon fibre production began in 1972 (Jones et al., 1982). The Task Group considered that the short duration of exposure in this study was insufficient for a reliable assessment of the potential health effects of these fibres.

Troitskaya (1988) reported that in a population of 327 examined workers in a PAN-based carbon fibre production facility, 67.9% had pharyngitis or rhinopharyngitis, 34% reported bronchitis and 39.6% had dermal effects. In addition to carbon fibres, workers were exposed to other substances such as ammonia, acrylonitrile and "carbon oxides". It was not possible for the Task Group to assess the validity of these results based on data provided in the published account of the study.

Cases of dermatitis in two workers were reported during a clean-up operation at the site of an aircraft crash where carbon fibres were detected at concentrations of up to 7 fibres/ml. Additional details on the monitoring methods used in this study are presented in section 4.2 (Formisano, 1989).

7.2 Aramid fibres

There was no change in diffusing capacity in workers (n = 167) involved in polyester fibre processing, for which exposures to para-aramid fibres and sulfur dioxide were "low", when compared to those in a non-exposed control group (n = 142) (Pal et al., 1990). The Task Group noted that no firm conclusions could be drawn on the basis of this study due to the lack of an *appropriate* control group.

Reinhardt (1980) briefly reported the results of patch testing of panels of human volunteers to assess skin irritancy and sensitization. In these studies, which involved more than 100 individuals, there was no skin sensitization but some minimal skin irritation following dermal contact with Kevlar or Nomex fabrics (Reinhardt, 1980). It was reported that because these fibres, especially Kevlar, are stiff, there is a potential for causing abrasive skin irritation under restrictive contact.

8. EVALUATION OF HUMAN HEALTH RISKS

8.1 Exposure

Many factors determine the exposure levels and airborne fibre characteristics for the synthetic organic fibres considered in this document. Most important among these factors are: (1) the nominal diameter of the parent fibre and distribution about this nominal diameter; and (2) the tendency of some fibre types to form smaller diameter fibres or to liberate smaller diameter fibrils during processing. Fibres of concern for deposition in the bronchoalveolar region in humans are those less than approximately 3 μm in diameter. Limited data are available concerning the fraction of fibres smaller than 3 μm in diameter for most synthetic organic fibrous products, although nominal fibre diameters are reported to be generally greater than 5 μm. Para-aramid fibres have fine-curled fibrils of less than 1 μm in diameter which can break off during processing. Some polyolefin fibres are produced which also have nominal diameters of 0.1-2 μm. Although carbon and graphite fibres normally have nominal diameters greater than 5 μm, some data suggest that diameters are reduced during incineration or other burning such as might occur after an aircraft crash.

Only limited data are available concerning occupational exposures to synthetic organic fibres and virtually no data are available with respect to environmental fate, distribution and general population exposures. Sampling and analytical methods used for measuring synthetic organic fibre exposures are those normally used for asbestos and man-made mineral fibres. Although these methods may be suitable, little method validation has taken place for synthetic organic fibres.

Occupational exposure data summarized in section 4.1 generally demonstrate low-level fibre exposures in fibre production facilities. Exposure levels in carbon fibre production are reported to be generally less than 0.1 fibres/ml, although levels of approximately 0.3 fibres/ml have been recorded. Exposure levels in facilities producing para-aramid fibres typically are less than 0.1 fibres/ml, although levels of over 2 fibres/ml have been measured during subsequent processing. Spinning and weaving of para-aramid staple yarns produces higher exposures (average 0.18 to 0.55 fibres/ml). Exposure levels in polypropylene fibre production and use are generally less than 0.1 fibres/ml, although levels of 0.5 fibres/ml have been reported. Virtually no data exist for secondary fibre uses and applications. Several exposure studies have reported levels in mg/m^3. While these data may be useful in overall evaluation of total dust exposures in the workplace, gravimetric determinations of airborne dust levels are of very limited value with respect to assessment of organic fibre exposures.

While airborne synthetic organic fibre concentrations for fibre types considered in this document are generally lower than 0.5 fibres/ml in the occupational environment, primarily in production, the possibility for higher exposures in different applications and uses exists, particularly for those operations that vigorously disturb fibres and the fibre matrix. Additionally, applications may involve exposures to other hazardous substances in the workplace.

8.2 Health effects

The potential adverse effects for humans from inhalation exposure to synthetic organic fibres are the development of malignant and non-malignant respiratory diseases. Concern for these effects is based on the health evidence derived from exposure to other respirable and durable fibrous materials. Other effects of concern include contact dermatitis and skin irritation from dermal exposures.

There is limited information on the health effects of synthetic organic fibres in humans. The information that does exist is considered inadequate for assessment because of numerous shortcomings including the inability to evaluate chronic effects due to the relative short period (20 years) that the fibres under review have been in production and use. However, there is some suggestive evidence that exposure to carbon and aramid fibres can cause contact dermatitis and skin irritation.

Toxicological data on carbon fibres and polyolefins fibres are also limited. The available animal data on acute and short-term inhalation exposures to carbon and polypropylene fibres indicate minimal respiratory system toxicity. Information on the chronic and carcinogenic effects of these fibres via inhalation is unavailable. However, the lack of appropriate animal data does not reduce the concern for potential health effects associated with long-term exposures to these fibres.

The available animal studies on exposure to respirable para-aramid fibrils indicate that acute and short-term inhalation exposures at concentrations as high as 1300 fibres/ml induce minimal pulmonary toxicity in rats. However, the results of the only chronic inhalation study indicate that respirable para-aramid fibres caused lung fibrosis ($\geq$ 25 fibres/ml) and lung neoplasms ($\geq$ 100 fibres/ml) in the rat. On the basis of limited available data, a potential for fibrogenic and carcinogenic effects may exist from exposure to these synthetic organic fibres in the occupational environment. The potential health risk associated with exposure to these synthetic organic fibres in the general environment is unknown at this time, but is likely to be very low.

9. CONCLUSIONS AND RECOMMENDATIONS FOR PROTECTION OF HUMAN HEALTH

The data reviewed in this report support the conclusion that respirable, durable organic fibres are of potential health concern. The following actions are suggested for protection of human health.

1. To the maximum extent possible, the organic fibres that are produced should be *non-inspirable* or at least *non-respirable*. Respirable fibres should not be produced by splitting or abrading during subsequent processing, use and disposal.

2. If small-diameter respirable fibres are necessary for specific products or applications, these fibres should not be *biopersistent* or exhibit other toxic effects.

3. All fibres that are *respirable* and *biopersistent* must undergo testing for toxicity and carcinogenicity. Exposures to these fibres should be controlled to the same degree as that required for asbestos until data supporting a lesser degree of control become available. The available data suggest that para-aramid fibres fall within this category. Furthermore, other respirable organic fibres should be considered to fall within this category until data indicating a lesser degree of hazard become available.

4. Populations potentially exposed to respirable organic fibres should have their exposure monitored in order to evaluate exposure levels and the possible need for additional control measures.

5. Populations identified as being those most exposed to respirable organic fibres should be enrolled in preventive medicine programmes that focus on the respiratory system. These data should be reviewed periodically for any early signs of adverse health effects.

10. FURTHER RESEARCH

10.1 Sampling and analytical methods

Sampling and analytical methods for synthetic organic fibres have generally been adapted from methods that have been used for asbestos. These include the use of membrane filter samples with analyses by phase contrast optical microscopy or limited use of scanning and transmission electron microscopy. Further validation of these methods for synthetic organic fibres is necessary, with particular attention to: (1) possible effects of high electrostatic charges of organic fibres on sampling and analysis; (2) effects of sample preparation on the integrity of fibres; and (3) combined effects of fibre size and refractive index on visibility of organic fibres under phase contrast microscopy. These parameters could introduce a negative bias in air sample results.

In addition, methods for sampling and analysis of synthetic organic fibres in biological tissues need further development and evaluation.

10.2 Exposure measurement and characterization

Much more information is needed relative to levels and characteristics of exposure in plants producing and using synthetic organic fibres. Complete size distributions, with special attention to those fibres less than approximately 3 μm in diameter, are needed for synthetic organic fibres. While some data are available for the fibre-producing industries, very little information concerning respirable fibre exposure is available in industries using or applying these fibres.

Information concerning environmental releases of synthetic organic fibres or fibre concentrations in environmental media is also very limited. Data need to be collected concerning the environmental fate and distribution of these materials and resultant non-occupational exposures.

10.3 Human epidemiology

No reliable data exist concerning chronic effects of synthetic organic fibre exposures. Multi-centre studies are needed in order to develop adequately sized cohorts for epidemiological studies. Both cross-sectional and longitudinal studies of respiratory morbidity, cancer mortality, and cancer incidence are needed.

10.4 Toxicology studies

With the exception of para-aramid fibres and fibrils, no adequate toxicity data are available for other synthetic organic fibres. These data are badly needed. Emphasis must be placed on chronic inhalation studies using respirable fibres of these materials. These studies must use fibre sizes which are respirable in the animal species being used and at levels which are at or near the maximum tolerated doses. They should include studies of tissue burden in order to confirm the expected tissue doses. A much better understanding of those characteristics of synthetic organic fibres (e.g., particle charge, agglomeration) that could effect deposition is needed.

More data are needed concerning the biopersistence of synthetic organic fibres. The critical period cf lung residence necessary for development of adverse health effects remains to be determined. Better test materials for measuring biopersistence are needed.

REFERENCES

Ahmed M (1982) Polypropylene fibres - Science and technology. New York, Elsevier Scientific Publishing Co., Inc.

Bernstein DM, Mast R, Anderson R, Hesterberg TW, Musselman R, Kamstrup O, & Hadley J (in press) An experimental approach to the evaluation of biopersistence of respirable synthetic fibers and minerals. Environ Health Perspect.

Bjork N, Ekstrand K, & Ruyter IE (1986) Implant-fixed, dental bridges from carbon/graphite fibre reinforced poly(methyl methacrylate). Biomaterials, 7(1): 73-75.

Boatman ES, Covert D, Kalman D, Luchtel D, & Omenn GS (1988) Physical, morphological, and chemical studies of dusts derived from machining of composite-epoxy materials. Environ Res, 45: 242-255.

Brinkman OA & Müller KM (1989) What's new in intraperitoneal test on Kevlar (asbestos substitute)? Pathol Res Pract, 185: 412.

Brown JR & Power AJ (1982) Thermal degradation of aramids: Part 1 - Pyrolysis chromatography/mass spectrometry of poly(1,3-phenylene isophthalamide) and poly(1,4-phenylene terephthalamide). Polym Degrad Stab, 4(5): 379-392.

Buchanan DR (1984) Olefin fibres. In: Kirk-Othmer encyclopedia of chemical technology, 3rd ed. New York, Chichester, Brisbane, Toronto, John Wiley and Sons, pp 357-385.

Busch H, Holländer W, Leusen K, Schilhabel J, Trasser FJ, & Nefer L (1989) Aerosol formation during laser cutting of fibre reinforced plastics. J Aerosol Sci, 20(8): 1473-1476.

Chiao CC & Chiao TT (1982) Aramid fibres and composites. In: Handbook of composites. New York, Van Nostrand Reinhold Co., pp 272-317.

Dalquist BH (1984) Evaluation of health aspects of carbon fibres. Chapel Hill, North Carolina, University of North Carolina, School of Public Health, Department of Environmental Sciences and Engineering, pp 21-22 (M. Sc. Thesis).

Davis JMG, Bolton RE, Dougglas AN, Jones AD, & Smith T (1988) Effects of electrostatic charge on the pathogenicity of chrysotile asbestos. Br J Ind Med, 45: 292-299.

Davis JMG (1987) Carcinogenicity of Kevlar aramid pulp following intraperitoneal injection into rats. Edinburgh, Institute of Occupational Medicine (Report No. PB88-109004).

Delmonte J (1981) Technology of carbon and graphite fiber composites. New York, Van Nostrand Reinhold Co.

DePass LR (1982) Evaluation of the dermal carcinogenicity of four carbon fiber materials in male C3H/HCJ mice. Export, Pennsylvania, Bushy Run Research Center.

Dunnigan J, Nadeau D, & Paradis D (1984) Cytotoxic effects of aramid fibres on rat pulmonary macrophages: comparison with chrysotile asbestos. Toxicol Lett, 20(3): 277-282.

Fevolden S & Moller M (1978) Lack of mutagenicity of extracts from polyethylene (PE) in strains of *Salmonella typhimurium*. Mutat Res, 53: 186-187.

Formisano JA (1989) Composite fibre field study: Evaluation of potential personnel exposures to carbon fibres during investigation of a military aircraft crash site. Appl Ind Hyg, 12/89(Spec Issue): 54-56.

Galli E (1981) Aramid fibres. Plast Compd, 4(6): 21.

Gieseke JA, Reif RB, & Schmidt EW (1984) Characterization of carbon fiber emissions from current and projected activities for the manufacture and disposal of carbon fiber products. Research Triangle Park, North Carolina, US Environmental Protection Agency (EPA-600/3-84-021).

Hanson MP (1980) Feasibility of Kevlar 49/Pmr-15 polyimide for high temperature applications. Cleveland, Ohio, National Aeronautics and Space Administration, Lewis Research Center (Report No. NASA-TM-81560; E-521).

Hartshorne AW & Laing DK (1984) The identification of polyolefin fibres by infrared spectroscopy and melting point determination. Forensic Sci Int, 26(1): 45-52.

Henderson JD, Mullarky RH, & Ryan DE (1987) Tissue biocompatibility of Kevlar aramid fibers and polymethylmethacrylate composites in rabbits. J Biomed Mater Res, 21(1): 59-64.

Henry WM, Melton CM, & Schmidt EW (1982) Method for measuring carbon fibre emissions from stationary sources. Washington, DC, US Environmental Protection Agency (EPA-600/3-82-080).

Hesterberg TW, Miller W, Christensen D, Law B, Hart G, Bunn W, & Anderson R (1991) Work place exposure and toxicology studies of polypropylene fibres. Littleton, Colorado, Schuller International, Inc.

Hesterberg TW, McConnell EE, Miller WC, Hamilton R, & Bunn WB (1992) Pulmonary toxicity of inhaled polypropylene fibres in rats. Fundam Appl Toxicol, 19(3): 358-366.

Hodgson AA (1989) The alternative raw materials. In: Hodgson AA ed. Alternatives to asbestos - the pros and cons. New York, Chichester, Brisbane, Toronto, John Wiley and Sons, pp 18-36 (Critical Reports on Applied Chemistry, Volume 26).

Hoffman AS (1977) Medical applications of polymeric fibers. Appl Polym Symp, 31: 313-334.

Holt PF (1982) Submicron dust in the guinea pig lung. Environ Res, 28: 434-442.

Holt PF & Horne M (1978) Dust from carbon fiber. Environ Res, 17: 276-283.

ICF Inc. (1986) Durable fibre exposure assessment. Washington, DC, US Environmental Protection Agency, Office of Pesticides and Toxic Substances.

ILO (1989) Safety in the use of mineral and synthetic fibres. Working document prepared for the Meeting of Experts on Safety in the Use of Mineral and Synthetic Fibres. Geneva, International Labour Office.

Jones HD, Jones TR, & Lyle WH (1982) Carbon fibre: results of a survey of process workers and their environments in a factory producing continuous filament. Ann Occup Hyg, 26(1-4): 861-868.

Kauffer JC, Vigneron JC, & Veissières U (undated) Emission de fibres lors de l'usinage de matériaux composites. Vandoeuvre, France, Département Environnement chimique, Promotion et Coordination des Actions d'Assistance.

Larsen G (1989) Experimental data on *in vitro* fibre solubility. In: Bignon J, Peto J, & Saracci R ed. Non-occupational exposure to mineral fibres. Lyon, International Agency for Research on Cancer, pp 134-139 (IARC Scientific Publications No. 90).

Law B, Bunn WB, & Hesterberg TW (1990) Solubility of polymeric organic fibers and man-made vitrous fibers in Gambles solution. Inhal Toxicol, 2: 321-339.

Lee KP, Kelly DP, & Kennedy GL (1983) Pulmonary response to inhaled Kevlar aramid synthetic fibers in rats. Toxicol Appl Pharmacol, 71(2): 242-253.

Lee KP, Trochimowicz HJ, & Reinhardt CF (1985) Pulmonary response of rats exposed to titanium dioxide (TiO_2) by inhalation for two years. Toxicol Appl Pharmacol, 79(2): 179-192.

Lee KP, Kelly DP, O'Neal FO, Stadler JC, & Kennedy GL (1988) Lung response to ultrafine Kevlar aramid synthetic fibrils following 2-year inhalation exposure in rats. Fundam Appl Toxicol, 11(1): 1-20.

Luchtel DL, Martin TR, & Boatman ES (1989) Response of the rat lung to respirable fractions of composite fiber-epoxy dusts. Environ Res, 48: 57-69.

Lurker PA & Speer RM (1984) Industrial hygiene evaluation of carbon composites machining. Presented at the Meeting on Health Effects of Carbon Composites, 26 January 1984. Dayton, Ohio, Wright Patterson Air Force Base.

McClellan RO, Miller FJ, Hesterberg TW, Warheit DB, Bunn WB, Dement JM, Kane AB, Lippmann M, Mast RW, McConnell EE, & Reinhardt CF (1992) Approaches to evaluating the toxicity and carcinogenicity of man-made fibers. Regul Toxicol Pharmacol, 16(3): 321-364.

Mahar S (1990) Particulate exposures resulting from the investigation and remediation of a crash site of an aircraft containing carbon composites. Am Ind Hyg Assoc J, 51: 459-481.

Makisato S, Paavolainen P, Holmstrom T, & Slatis P (1984) Tissue reaction after implantation of carbon fibre and polypropylene. An experiment on rats. Acta Orthop Scand, 55: 675-676 (Abstract).

Maltoni C & Minardi F (1989) Recent results of carcinogenicity bioassays of fibres and other particulate materials. In: Bignon J, Peto J, & Saracci, R ed. Non-occupational exposure to mineral fibres. Lyon, International Agency for Research on Cancer, pp 46-53 (IARC Scientific Publications No. 90).

Martin TR, Meyer SW, & Luchtel DR (1989) An evaluation of carbon fiber composites for lung cells *in vitro* and *in vivo*. Environ Res, 49: 246-261.

M.B. Research Laboratories (1980) Twenty-one month observation after single intratracheal insufflation (Project No. 78-2806). Research Triangle Park, North Carolina, US Environmental Protection Agency.

Merriman EA (1992) A safety-in-use program for para-aramid fibre. Presented at the American Industrial Hygiene Conference and Exposition, 5 June 1992. (Paper No. 232).

Moss CE & Seitz T (1990) Hazard evaluation and technical assistance report. Cincinnati, Ohio, National Institute of Occupational Safety and Health (NIOSH HETA 90-102-L2075).

Neugebauer R, Helbing G, Wolter O, Mohr W, & Gistinger G (1981) The body reaction to carbon fiber particles implanted into the medullary space of rabbits. Biomaterials, 2: 182-184.

NIOSH (1989) Manual of analytical methods. Method 7400 (Revision 3), 3rd ed. Cincinnati, Ohio, National Institute of Occupational Safety and Health (DHHS (NIOSH) Publication No 84-100).

Office of Metals, Minerals and Commodities, Nonferrous Metals Division (1985) A competitive assessment of selected reinforced composite fibers. Washington, DC, US Department of Commerce, International Trade Administration (ITA).

Owen PE, Glaister JR, Ballantyne B, & Clary JJ (1986) Subchronic inhalation toxicology of carbon fibers. J Occup Med, 28(5): 373-376.

Pal TM, Schaaphok J, Coenraads J (1990) Lung function of workers occupied in drawing and treatment of aramid fibres. Cah Notes Doc, 138: 254-257.

Parsons JR, Bhayani S, Alexander H, & Weiss AB (1985) Carbon fibre debris within the synovial joint. Clin Orthop Related Res, 196: 69-76.

Pott F, Ziem U, Reiffer FJ, Huth F, Ernst H, & Mohr U (1987) Carcinogenicity studies on fibres, metal compounds, and some other dusts in rats. Exp Pathol, 32: 129-152.

Pott F, Roller M, Ziem U, Reiffer FJ, Bellmann B, Rosenbruch M, & Huth F (1989) Carcinogenicity studies on natural and man-made fibres with the intraperitoneal test in rats. In: Bignon J, Peto J, & Saracci R ed. Non-occupational exposure to mineral fibres. Lyon, International Agency for Research on Cancer, pp 173-179 (IARC Scientific Publications No. 90).

Pott F, Dungworth DL, Heinrich U, Muhle H, Kamino K, Germann PG, Roller M, Rippe RM, & Mohr U (in press) Lung tumours in rats after intratracheal instillation of dusts. Annals of Occupational Hygiene.

Preston J (1979) Aramid fibres. In: Kirk-Othmer encyclopedia of chemical technology, 3rd ed. New York, Chichester, Brisbane, Toronto, John Wiley and Sons, pp 213-242.

Razinet ML, Dipietro C, & Merrit C (1976) Thermal degradation studies of fibres and composite base materials. Watertown, Massachusetts, Army Materials and Mechanics Research Center.

Reinhardt CF (1980) Toxicology of aramid fibres. In: Proceedings of the National Workshop on Substitutes for Asbestos. Washington, DC, US Environmental Protection Agency, pp 443-449 (EPA-560/3-80-001).

SRI International (1985) Chemical economics handbook. Menlo Park, California, SRI International, section 543.5811A.

SRI International (1987) Chemical economics handbook. Menlo Park, California, SRI International, section 543.5811A.

Styles JA & Wilson J (1973) Comparison between *in vitro* toxicity of polymer and mineral dusts and their fibrogenicity. Ann Occup Hyg, **16**: 214-250.

Sussholz B (1980) Environmental exposure considerations to the release of graphite fibers during aircraft fires In: Proceedings of the National Workshop on Substitutes for Asbestos. Washington, DC, US Environmental Protection Agency, pp 451-474 (EPA-560/3-80-001).

Tayton K, Phillips G, & Ralis Z (1982) Long term effects of carbon fiber on soft tissues. J Bone Jt Surg, **64**: 112.

Troitskaya NA (1988) [Hygienic evaluation of working conditions in the polyacrylonitrile-based production of carbon fiber.] Gig i Sanit, **4**: 21-23 (in Russian with English summary).

US EPA (1980) Air quality criteria for sulfur oxides and particulates. Washington, DC, US Environmental Protection Agency, pp 11/28-11/32 (EPA-600/8-82-029c).

US EPA (1988) Health hazard assessment of nonasbestos fibers. Research Triangle Park, North Carolina, US Environmental Protection Agency (Report No. OPTS-62036).

Verwijst LPF (undated) Measuring exposure to fibres at the workplace design and implementation of a monitoring system. Arnheim, The Netherlands, Akzo NV.

Volk HF (1979) Carbon (carbon and artificial graphite). In: Grayson M & Eckroth D ed. Kirk-Othmer encyclopedia of chemical technology, 3rd ed. New York, Chichester, Brisbane, Toronto, John Wiley and Sons, pp 556-628.

Wagman J, Berger H, Miller J, & Conner W (1979) Dust and residue from machining and incinerating graphite/epoxy composites. A preliminary study. Washington, DC, Government Printing Office.

Warheit DB, Kellar KA, & Hartsky MA (1992) Pulmonary cellular effects in rats following aerosol exposures to ultrafine Kevlar aramid fibrils: evidence for biodegradability of inhaled fibrils. Toxicol Appl Pharmacol, **116**: 225-239.

Warheit DB, Hansen JF, Carakostas MC, & Hartsky MA (in press/a) Acute inhalation toxicity studies in rats with a respirable-sized experimental carbon fiber: pulmonary biochemical and cellular effects. Ann Occup Hyg.

Warheit DB, Hartsky MA, McHugh TA, & Kellar KA (in press/b) Biopersistence of inhaled organic and inorganic fibers in the lungs of rats. Environ Health Perspect.

WHO (1985) Reference methods for measuring airborne man-made mineral fibres. Copenhagen, World Health Organization, Regional Office for Europe (Environmental Health Report 4).

WHO (1988) IPCS Environmental Health Criteria 77: Man-made mineral fibres. Geneva, World Health Organization, 165 pp.

APPENDIX 1. SUMMARY OF PATHOLOGY WORKSHOP ON THE LUNG EFFECTS OF PARA-ARAMID FIBRILS AND TITANIUM DIOXIDE
(5-6 October 1992)

An international panel of 13 pathologists met to evaluate the cystic lesion observed in the lungs of rats in chronic inhalation studies with Kevlar[1] aramid fibrils (Lee et al., 1988) and titanium dioxide particles (Lee et al., 1985). Slides representative of the entire spectrum of cystic keratinizing lesions observed in these studies were sent to each of these pathologists prior to the meeting.

The pathologists agreed that the most appropriate morphologic diagnosis is Proliferative Keratin Cyst (PKC). The lesion is a cyst lined by a well-differentiated stratified squamous epithelium and with a central keratin mass. Growth appears to have occurred by keratin accumulation and by peripheral extension into the alveolar spaces. The lesion is sharply demarcated except in those areas in which there has been extension into adjacent alveoli. The squamous epithelium has few mitotic figures and dysplasia is absent.

All participants agreed that the lesion is not a malignant neoplasm. The majority was of the opinion that the lesion is not neoplastic. A minority considered that the lesion is probably a benign tumour. The participants had not seen a similar lung lesion in humans.

Morphologic features of the cystic lesion that were used to exclude malignancy were the lack of invasion of the pleura, blood vessels or the mediastinum as well as the absence of dysplasia and the paucity of mitotic figures.

---

[1] Kevlar is registered trademark of the DuPont Company for its para-aramid fibre.

PARTICIPANTS IN PATHOLOGY WORKSHOP ON THE LUNG EFFECTS OF PARA-ARAMID FIBRILS AND TITANIUM DIOXIDE
(5-6 October 1992)

Participants

Dr M. Brockmann, Institut für Pathologie, Universitätsklinik "Bergmann's Heil", Bochum, Germany

Dr W.W. Carlton, Department of Veterinary Pathobiology, School of Veterinary Medicine, Purdue University, West Lafayette, Indiana, USA

Dr J.M.G. Davis, Institute of Occupational Medicine, Edinburgh, United Kingdom

Dr V.J. Feron, T.N.O. Toxicology and Nutrition Institute, Zeist, Netherlands

Dr M. Kuschner, Pathology Department, SUNY at Stonybrook Health Science Center, Stonybrook, New York, USA

Dr K.P. Lee, Haskell Laboratory, E.I. du Pont de Nemours & Company, Haskell Laboratory for Toxicology and Industrial Medicine, Newark, Delaware, USA

Dr L.S. Levy, Institute of Occupational Health, University of Birmingham, Birmingham, United Kingdom (*Chairman*)

Dr E. McCaughey, Canadian Reference Center for Cancer Pathology, Ottawa Civic Hospital, Ottawa, Ontario, Canada

Dr P. Nettesheim, Laboratory of Pulmonary Pathobiology, National Institute of Environmental Health Sciences, Research Triangle Park, North Carolina, USA

Dr K. Nikula, Inhalation Toxicology Research Institute, Albuquerque, New Mexico, USA

Dr R. Renne, Batelle Pacific Northwest, Richland, Washington, USA

Dr M. Schultz, Institut für Pathologie, Bezirkskrankenhaus Magdeburg, Magdeburg, Germany

Dr V.S. Turusov, Cancer Research Center, Russian Academy of Medical Sciences, Moscow, Russia

Dr J.C. Wagner, Preston, Weymouth, Dorset, United Kingdom

Observers

Dr C.L.J. Braun, Akzo NV, Arnhem, Netherlands

Dr R.C. Brown, MRC Toxicology Unit, Medical Research Council Laboratories, Carshalton, Surrey, United Kingdom

Dr S.R. Frame, E.I. du Pont de Nemours & Company, Haskell Laboratory for Toxicology and Industrial Medicine, Newark, Delaware, USA

Dr N.F. Johnson, Inhalation Toxicology Research Institute, Albuquerque, New Mexico, USA

Dr G.L. Kennedy, Jr, E.I. du Pont de Nemours & Company, Haskell Laboratory for Toxicology and Industrial Medicine, Newark, Delaware, USA

Dr E.A. Merriman, E.I. du Pont de Nemours & Company, Wilmington, Delaware, USA

Dr C.F. Reinhardt, E.I. du Pont de Nemours & Company, Haskell Laboratory for Toxicology and Industrial Medicine, Newark, Delaware, USA

Dr J.W. Rothuizen, Rothuizen Consulting, En Thiéré, Genolier, Switzerland

Dr O. von Susani, Du Pont International SA, Grand-Saconnex, Geneva, Switzerland

Dr D.B. Warheit, E.I. du Pont de Nemours & Company, Haskell Laboratory for Toxicology and Industrial Medicine, Newark, Delaware, USA

RESUME

1. Identité, propriétés physiques et chimiques

Les fibres de carbone/graphite sont des filaments de carbone produits par traitement à haute température de l'une ou l'autre des trois matières premières suivantes: rayonne (cellulose régénérée), brais de goudron, de houille ou de pétrole, or encore polyacrylonitrile (PAN). Le diamètre nominal des fibres de carbone varie de 5 à 15 μm. Les fibres de carbone sont flexibles; elles conduisent l'électricité et la chaleur et les variétés à haute performance sont dotées d'un module de Young élevé (coefficient d'élasticité qui mesure la souplesse ou la rigidité d'un matériau) et d'une forte résistance à la traction. Elles sont résistantes à la corrosion, légères, réfringentes et chimiquement inertes (sauf à l'oxydation). En outre, elles sont très stables à la traction, possèdent un faible coefficient de dilatation thermique et une faible densité et sont très résistantes à l'abrasion et à l'usure.

Les fibres aramides sont préparées par réaction de diamines aromatiques sur des dichlorures d'acides aromatiques. On les produit sous la forme de filaments continus, de fibres discontinues et de pulpe. Il existe deux types principaux de fibres aramides, le para- et le méta-aramide, qui ont toutes deux un diamètre nominal de 12 à 15 μm. Les fibres para-aramides sont parfois munies de fibrilles finement recourbées et enchevêtrées qui se situent dans la gamme des particules respirables (< 1 μm de diamètre) sur la surface de la fibre centrale. Ces fibrilles peuvent se détacher par abrasion lors de la fabrication ou de l'utilisation des fibres et être suspendus dans l'atmosphère. En général, les fibres aramides présentent une résistance à la traction moyenne à très élevée, une résistance à l'allongement moyenne et un module de Young moyen à très élevé. Elles résistent à la chaleur, aux produits chimiques et à l'abrasion.

Les fibres polyoléfiniques sont constituées de polymères à longue chaîne composés d'au moins 85% en poids d'unités d'éthylène, de propylène et d'autres oléfines; le polyéthylène et le polypropylène sont utilisés dans le commerce. A part pour certains types comme les microfibres, le diamètre nominal de la plupart des fibres de polyoléfine est suffisamment important pour que très peu d'entre elles se situent dans la limite des particules respirables.

Les polyoléfines sont très hydrophobes et chimiquement inertes. Leur résistance à la traction est beaucoup plus faible que celle des fibres de carbone ou des fibres d'aramide et elles sont relativement inflammables, avec un point de fusion situé entre 100 et 200 °C.

Les méthodes qui ont été mises au point pour le comptage des fibres minérales sont utilisées en hygiène industrielle pour la surveillance des fibres organiques synthétiques. Toutefois la valeur de ces méthodes n'a pas été vérifiée dans ce cas. Des facteurs tenant aux propriétés électrostatiques, à la solubilité dans la préparation des échantillons et à l'indice de réfraction peuvent poser des problèmes lorsqu'on a recours à ces méthodes.

2. Sources d'exposition humaine et environnementale

On estime que la production mondiale de fibres de carbone et de graphite a dépassé 4000 tonnes en 1984. En ce qui concerne les fibres aramides, elle était supérieure à 30 000 tonnes en 1989 et la production de fibres polyoléfiniques dépassait les 182 000 tonnes pour les seuls Etats-Unis d'Amérique. Les fibres de carbone et d'aramide sont utilisées principalement pour la fabrication de matériaux composites dont les industries aérospatiales, militaires et autres ont besoin pour améliorer la résistance, la rigidité, la durabilité et la conductivité électrique ou la tenue à la chaleur de certains éléments. Les fibres polyoléfiniques sont surtout utilisées dans l'industrie textile.

Des cas d'exposition aux fibres organiques synthétiques sur les lieux de travail ont été rapportés. Ces fibres peuvent pénétrer dans l'environnement lors de la production, de la transformation ou de la combustion des matériaux composites ou lors de leur mise au rebut. On ne dispose que de très peu de données sur les rejets effectifs de ces produits dans l'environnement.

Les données dont on dispose sur le transport, la distribution et la transformation des fibres organiques dans l'environnement se limitent à l'identification des produits résultant de l'incinération, par les services municipaux, des déchets de matériaux composites à base de fibres de carbone et des produits de pyrolyse des fibres de carbone et des fibres aramides. Des expériences au cours desquelles on a simulé la combustion dans des incinérateurs municipaux ont permis de constater qu'il y avait réduction du diamètre et de la longueur des fibres de carbone. La décomposition pyrolytique de fibres de carbone et d'aramide produit principalement des hydrocarbures aromatiques, du dioxyde et de monoxyde de carbone et des cyanures.

3. Concentrations dans l'environnement et exposition de l'homme

Des poussières de fibres organiques synthétiques peuvent être libérées sur les lieux de travail lors d'opérations telles que la production, le bobinage, coupe le tissage et la coupe des fibres, de même qu'au cours de l'usinage, de la réalisation et de la manipulation des matériaux composites.

Dans le cas des fibres de carbone et de graphite, les concentrations en fibres respirables sont généralement inférieures à 0,1 fibres/ml, mais on a mesuré des concentrations pouvant aller jusqu'à 0,3 fibres/ml à proximité des opérations de coupe ou de bobinage. Des fibres peuvent également être libérées dans l'environnement lors de l'usinage (perçage, sciage, etc) des composites à base de fibres de carbone, encore que la plupart des matériaux respirables ainsi produits soient non fibreux.

On a fait état, sur les lieux de travail, de concentrations moyennes en fibrilles de para-aramide aéroportées inférieures à 0,1 fibrille/ml au cours des opérations concernant les fils continus et à 0,2 fibrilles/ml lors de la coupe en floc et en pulpe. En filature, on observe des concentrations moyennes en fibrilles aéroportées qui se caractérisent par une valeur inférieure à 0,5 fibrille/ml, mais on a aussi fait état de concentrations de l'ordre de 2,0 fibrilles/ml. On note aussi des valeurs caractéristiques de l'exposition moyenne inférieures à 0,1 fibrille/ml pour d'autres applications avec des maxima atteignant 0,3 fibrilles/ml. On a mis en évidence un risque particulier d'exposition lors du découpage des composites au jet hyperbarique, les concentrations pouvant atteindre 2,91 fibrilles/ml. Au cours de découpage au laser de résines époxy renforcées par des fibres d'aramide, des particules d'un diamètre aérodynamique moyen de 0,21 μm on été observées mais l'étude en question n'indique pas la teneur en fibrilles de la poussière. Au cours de ce type d'opérations, il y également production d'un certain nombre de composés organiques volatils (notamment du benzène, du toluène, du benzonitrile et du styrène) ainsi que des gaz comme le cyanure d'hydrogène, le monoxyde de carbone et le dioxyde d'azote.

D'après des données limitées établies durant la surveillance de l'air dans un atelier produisant des fibres de polypropylène, les concentrations maximales de fibres aéroportées d'une longueur dépassant les 5 μm, se situent autour de 0,5 fibre/ml, la plupart des valeurs étant inférieures à 0,1 fibre/ml. L'examen au microscope électronique à balayage a montré que le diamètre des fibres aéroportées allait de 0,25 à 3,5 μm et leur longueur de 1,7 à 69 μm. Dans un échantillon unique d'air ambiant prélevé à proximité d'une unité de tissage de fibres de carbone, on a relevé une concentration de 0,0003 fibre/ml. Les dimensions moyennes des fibres étaient de 706 μm par 3,9 μm. On a également signalé la présence de fibres de carbone à l'endroit où s'étaient écrasés deux aéronefs militaires, présence attribuable à la combustion du composite de fibres de carbone utilisé pour la construction de ces appareils. Aucune autre donnée utile sur les concentrations dans l'environnement n'a été fournie.

4. Dépôt, élimination, rétention, persistance et redistribution

On ne dispose que de peu de données sur les fibres organiques synthétiques. Des données relatives aux fibres de para-aramide (Kevlar) indiquent que ces fibres se déposent au niveau de la bifurcation des canaux alvéolaires. On pense également que ces fibres sont transportées jusqu'aux ganglions lymphatiques trachéo-bronchiques.

5. Effets sur les animaux de laboratoire et les systèmes *in vitro*

On manque de données satisfaisantes résultant d'études expérimentales valables sur les divers types de fibres organiques synthétiques examinées ici.

Aucune étude valable n'a été consacrée à l'examen du pouvoir fibrogène ou cancérogène des fibres de carbone et de graphite. Parmi les effets observés à la suite de l'exposition, pendant quelques jours, par la voie respiratoire, de rats à des fibres de dimension respirable produites à partir de différents types de brais, on a relevé des réactions inflammatoires, un accroissement du remplacement des cellules parenchymateuses et une hyperplasie minime des cellules alvéolaires (pneumocytes de Type II). Les données fournies par une étude au cours de laquelle on a procédé à des instillations intratrachéennes et à des injections intrapéritonéales, sont jugées insuffisantes pour permettre une évaluation, du fait que les matériaux étudiés n'ont pas été caractérisés et que l'étude ne donne pas suffisamment de renseignements sur le protocole suivi et les résultats obtenus. Une étude par badigeonnage cutané sur la souris, portant sur quatre types de fibres de carbone en suspension dans le benzène, a été considérée comme insuffisante pour permettre une évaluation de l'activité cancérogène de ces fibres.

Dans le cas des fibres de para-aramide, l'essentiel des données résulte de l'expérimentation sur le Kevlar. Des études de brève durée (deux semaines), au cours desquelles des animaux ont été exposés à de la poussière de Kevlar par la voie respiratoire, ont montré qu'il y avait une réaction dont l'intensité diminuait après cessation de l'exposition au niveau des macrophages pulmonaires. Des études de brève durée portant sur des fibrilles ultrafines de Kevlar, ont révélé une réaction analogue au niveau de macrophages avec des plaques d'épaississement au niveau des canaux alvéolaires. Ces deux types de lésion ont également régressé après l'exposition, mais trois à six mois plus tard on observait encore une fibrose minime résiduelle. L'exposition par la voie respiratoire, pendant deux ans, de rats à des fibrilles de Kevlar a produit une fibrose pulmonaire liée à cette exposition (à une concentration > 25 fibres/ml) et a conduit à la formation de tumeurs pulmonaires (11% à la concentration de 400 fibres/ml et 6% à la concentration de 100 fibres/ml chez les femelles; 3% à la concentration de 400 fibres/ml chez les mâles) d'un type inhabituel (carcinomes spino-cellulaires kystiques kératinisants). Une surmortalité due à une toxicité pulmonaire a été observée à la concentration la plus élevée, ce qui indique que la dose maximale tolérable avait été dépassée. La portée biologique de ces lésions et leur signification pour la santé humaine ont été très débattues. Il est possible que cette étude, qui s'est achevée au bout de 24 mois, n'ait pas révélé la totalité du pouvoir cancérogène des fibrilles.

L'instillation intertrachéenne d'une dose unique de papier désagrégé Nomex (2,5 mg) contenant des fibres d'un diamètre allant de 2 à 30 μm a produit une réaction inflammatoire non spécifique. Une réaction granulomateuse s'est produite deux ans après l'exposition. L'instillation intratrachéenne d'une dose unique de 25 mg de Kevlar a provoqué une réaction inflammatoire non spécifique qui a disparu en l'espace d'environ une semaine. Ultérieurement, on a observé une réaction granulomateuse et une fibrose minime.

Dans trois études, l'injection intrapéritonéale de fibres de Kevlar (jusqu'à 25 mg/kg) a induit une réponse granulomateuse sans toutefois que l'incidence des néoplasmes ne présente d'augmentation significative. Selon les auteurs de cette étude, l'absence de réponse tumorale pourrait s'expliquer par l'agglomération des fibrilles de Kevlar dans la cavité péritonéale.

Il n'existe pas d'études valables au cours desquelles on ait examiné le pouvoir cancérogène ou fibrogène des fibres de polyoléfines. Une étude au cours de laquelle on a exposé des rats à des fibres respirables de polypropylène pendant 40 jours par la voie respiratoire (46% des fibres < 1 μm) à des concentrations allant jusqu'à 50 fibres/ml, a permis de constater des modifications liées à la dose et à la durée de l'exposition et qui se caractérisaient par un accroissement de la cellularité et une bronchiolite. On ne dispose d'aucune donnée utile sur l'effet de l'instillation intratrachéenne. Des études au cours desquelles on a injecté à des rats des fibres ou de la poussière de polypropylène dans la cavité intrapéritonéale, n'ont pas révélé d'accroissement sensible du nombre de tumeurs péritonéales.

On ne dispose pas de données suffisantes pour pouvoir évaluer la toxicité *in vitro* et la génotoxicité des fibres organiques synthétiques. Dans le cas des aramides, les études montrent que le fibrilles de para-aramide fines et courtes présentent des propriétés cytotoxiques. Pour ce qui est des fibres polyoléfiniques, il semblerait que les fibres de polypropylène présentent une certaine cytotoxicité. Les tests de mutagénicité sur des extraits de granulés de polyéthylène n'ont donné que des résultats négatifs.

6. Effets sur l'homme

Une étude transversale relative à 88 des 110 employés d'une unité de production de fibres de carbone continues à base de polyacrylonitrile, n'a révélé aucun effet nocif sur la fonction respiratoire comme l'ont montré les examens radiographiques et spirométriques et les questionnaires sur les symptômes respiratoires. D'autres études moins bien documentées ont fait état d'effets indésirables chez des ouvriers travaillant à la production de fibres de carbone et de polyamide; les données qui figurent dans ces publications sont toutefois insuffisantes pour qu'on puisse se prononcer sur la validité des inférences indiquées.

7. Résumé de l'évaluation

On ne dispose que de données limitées sur les niveaux d'exposition à la plupart des fibres organiques synthétiques. Lʳ données disponibles indiquent en général que sur les lieux de travail, l'exposition est faible. Toutefois il subsiste un risque d'exposition plus importante lors d'applications et d'utilisations futures. On ne dispose pratiquement d'aucune donnée sur la destinée et la répartition dans l'environnement de ces fibres ni sur l'exposition de la population générale.

En se basant sur les données toxicologiques limitées fournies par l'expérimentation animale, on peut conclure qu'il existe une possibilité d'effets nocifs sur la santé en cas d'exposition par la voie respiratoire à ces fibres organiques de synthèse sur le lieu de travail. On ignore pour l'instant le risque qu'impliquerait l'exposition à ces fibres dans l'environnement général, mais il est probablement très faible.

RESUMEN

1. Identidad, propiedades físicas y químicas

Las fibras de carbono/grafito son formas filamentosas de carbón que se obtienen procesando a alta temperatura alguno de los tres materiales precursores siguientes: rayón (celulosa regenerada), brea (residuo de petróleo o alquitrán) o poliacrilonitrilo (PAN). El diámetro nominal de las fibras de carbono oscila entre 5 y 15 μm. Las fibras de carbono son flexibles, eléctrica y térmicamente conductivas, y sus variedades de alto rendimiento poseen un módulo de Young (coeficiente de elasticidad que refleja la mayor o menor rigidez del material) alto y una gran resistencia a la tracción. Son resistentes a la corrosión, ligeras, refractivas y químicamente inertes (excepto a la oxidación), y presentan una gran estabilidad frente a las fuerzas de tracción, una baja densidad y expansión térmica y una alta resistencia a la abrasión y al desgaste.

Las fibras de aramida se forman por reacción entre diaminas aromáticas y dicloroácidos aromáticos. Son producidas en forma de filamentos continuos, hebras y pulpa. Hay dos tipos principales de fibras de aramida: para- y meta-aramida, ambas con un diámetro nominal de 12-15 μm. Las fibras de para-aramida pueden presentar, adheridas a la superficie de su parte central, fibrillas muy rizadas y entrelazadas del tamaño de las partículas respirables (< 1 μm de diámetro). Estas fibrillas pueden desprenderse y quedar suspendidas en el aire en caso de abrasión durante su fabricación o empleo. Por lo general, las fibras de aramida presentan una resistencia a la tracción entre mediana y muy alta, una elongación entre mediana y baja, y un módulo de Young entre moderado y muy alto. Son resistentes al calor, a los productos químicos y a la abrasión.

Las fibras de poliolefina son polímeros de cadena larga compuestos por al menos un 85% (respecto al peso) de etileno, propileno u otras unidades de olefina; el polietileno y el polipropileno se emplean comercialmente. Salvo algunas excepciones, como la microfibra, los diámetros nominales de la mayoría de los distintos tipos de fibras de poliolefina son bastante grandes, y no abundan los de tamaño respirable.

Las poliolefinas son extremadamente hidrofóbicas e inertes. Su resistencia a la tracción es notablemente inferior a la del carbono o de las fibras de aramida. Son relativamente inflamables, y sus temperaturas de fusión oscilan entre 100 y 200 °C.

Para la vigilancia, a efectos de higiene industrial, de las fibras orgánicas sintéticas se han utilizado métodos desarrollados para contar fibras minerales, métodos que, no obstante, no han sido validados para esa finalidad. Factores tales como las propiedades electrostáticas, la solubilidad en el líquido de montaje y el índice refractivo pueden ser fuente de problemas al emplear esos métodos.

2. Fuentes de exposición humana y ambiental

La producción mundial estimada de fibras de carbono y grafito superó las 4000 toneladas en 1984. Por lo que se refiere a la aramida, se superaron las 30 000 toneladas en 1989, y la producción de fibras de poliolefina sobrepasó las 182 000 toneladas (sólo en los Estados Unidos de América). Las fibras de carbono y aramida se emplean principalmente para fabricar materiales compuestos avanzados en las industrias aeroespacial, militar y otras, con el objeto de mejorar su resistencia, rigidez, durabilidad, conductividad eléctrica o resistencia térmica. Las fibras de poliolefina se utilizan normalmente en la industria textil.

Se han descrito casos de exposición a fibras orgánicas sintéticas en el medio ocupacional. Durante la producción, elaboración o combustión de materiales compuestos, así como durante su evacuación, puede producirse una liberación de fibras orgánicas sintéticas en el medio ambiente. Se dispone de muy pocos datos sobre la liberación real de esos materiales en el entorno.

Los datos disponibles sobre el transporte, distribución y transformación de fibras orgánicas en el medio ambiente se limitan a la identificación de los productos que resultan de la incineración municipal de los desechos a partir de compuestos que contienen fibras de carbono y de productos de la descomposición pirolítica de fibras de carbono y aramidas. Durante la simulación de la incineración municipal se redujeron tanto el diámetro como la longitud de las fibras de carbono. Entre los principales productos de la descomposición pirolítica de fibras de carbono y aramida figuran hidrocarburos aromáticos, dióxido de carbono, monóxido de carbono, y cianuros.

3. Niveles ambientales y exposición humana

En el lugar de trabajo se liberan fibrillas orgánicas sintéticas durante operaciones tales como la producción, bobinado, troceado, entrelazado, corte y maquinado de las fibras, así como durante la formación y manipulación de materiales compuestos.

En el caso de las fibras de carbono/grafito, las concentraciones de fibra respirable son por lo general inferiores a 0,1 fibras/ml, pero se han detectado concentraciones de hasta 0,3 fibras/ml en las proximidades de instalaciones de troceado o bobinado. También pueden liberarse fibras durante el maquinado (perforación, aserrado, etc.) de compuestos de fibra de carbono, si bien la mayor parte del material respirable generado en esos casos no es de carácter fibroso.

Se ha notificado que las concentraciones medias de fibrillas de para-aramida en suspensión en el aire del lugar de trabajo son inferiores a 0,1 fibrillas/ml trabajando filamentos, y de menos de 0,2 fibrillas/ml en las instalaciones de corte de vedijas y fabricación de pulpa. Durante el procesamiento de la hilaza, la concentración media de fibrillas en suspensión en el aire es generalmente inferior a 0,5 fibrillas/ml, si bien se han notificado niveles de hasta aproximadamente 2,0 fibrillas/ml. Otras exposiciones asociadas a usos finales en el lugar de trabajo son normalmente inferiores a 0,1 fibrillas/ml como promedio, cifrándose las exposiciones máximas en 0,3 fibrillas/ml. Se ha demostrado que conlleva un riesgo especial de exposición el corte de materiales compuestos por chorro de agua hiperbárico, operación en la que se han detectado niveles de hasta 2,91 fibrillas/ml. Se han generado partículas de un diámetro aerodinámico medio de 0,21 μm durante el corte mediante láser de epoxiplásticos reforzados con fibras de aramida, pero no se ha notificado el contenido de fibra del polvo. Durante esas operaciones se producen también determinados compuestos orgánicos volátiles (en particular benceno, tolueno, benzonitrilo y estireno) y otros gases (cianuro de hidrógeno, monóxido de carbono y dióxido de nitrógeno).

Los datos limitados obtenidos en una fábrica de producción de fibras de polipropileno muestran que en el caso de las fibras de más de 5 μm de longitud su concentración máxima en el aire era de 0,5 fibras/ml, siendo la mayoría de los valores inferiores a 0,1 fibras/ml. La microscopía electrónica de barrido mostró que el tamaño de las fibras suspendidas en el aire oscilaba entre

0,25 y 3,5 μm de diámetro y 1,7 y 69 μm de longitud. En una sola muestra ambiental recogida cerca de una tejeduría de fibra de carbono, se detectó una concentración de 0,0003 fibras/ml. La dimensión de esas fibras era como promedio de 706 μm por 3,9 μm. Se ha notificado también la liberación de fibra de carbono en el lugar de colisión de dos aeronaves militares, de resultas de la combustión del compuesto de fibra de carbono empleado en su construcción. No se obtuvo ningún otro dato de interés sobre las concentraciones presentes en el medio ambiente.

4. Depósito, eliminación, retención, durabilidad y translocación

La información obtenida acerca de fibras orgánicas sintéticas específicas es escasa. Los datos referentes a las fibras de para-aramida inhaladas (Kevlar) indican que éstas se depositan en las bifurcaciones de los conductos alveolares. Hay también indicios de translocación a los nódulos linfáticos traqueobronquiales.

5. Efectos en los animales de experimentación y en los sistemas de prueba *in vitro*

En lo que respecta a los tipos de fibra orgánica sintética aquí analizados, son muy pocos los datos de buena calidad aportados por los estudios experimentales realizados al efecto.

No hay ningún estudio en que se haya examinado adecuadamente el potencial fibrógeno o carcinógeno de las fibras de carbono/grafito. En ratas expuestas a la inhalación de corta duración (unos días) de fibras de tamaño respirable obtenidas a partir de brea se observaron respuestas inflamatorias, un aumento de la velocidad de recambio de las células parenquimatosas y una hiperplasia mínima de las células alveolares de tipo II. Se considera que los resultados de un estudio realizado mediante instilación intratraqueal e inyección intraperitoneal no se prestan a evaluación, debido a la insuficiente caracterización del material de ensayo y a la falta de documentación adecuada sobre el protocolo y los resultados. Un estudio realizado con ratones a los que se pintó la piel con cuatro tipos de fibra de carbono suspendidos en benceno resultó inadecuado para evaluar la actividad carcinógena.

En el caso de las fibras de paraaramida, la mayor parte de los datos proceden de experimentos realizados con Kevlar. Los estudios efectuados sobre los efectos de la inhalación de corta duración (2 semanas) de polvo de Kevlar han puesto de manifiesto una respuesta macrofágica pulmonar cuya gravedad disminuye tras la interrupción de la exposición. Los estudios de corta duración realizados con fibrillas de Kevlar ultrafinas han revelado una respuesta macrofágica parecida y un espesamiento desigual de los conductos alveolares. Esos dos tipos de lesiones remitieron también tras la exposición, pero a los 3-6 meses persistía todavía un grado mínimo de fibrosis. En un estudio realizado con ratas a las que se sometió durante dos años a inhalación de fibrillas de Kevlar se observó la aparición, relacionada con esa exposición, de fibrosis pulmonar (a concentraciones superiores a 25 fibras/ml) y de neoplasias pulmonares de un tipo inusitado (carcinoma escamoso quístico queratinizante) en un 11% de hembras a concentraciones de 400 fibras/ml, en un 6% de hembras a concentraciones de 100 fibras/ml; y en un 3% en los machos a concentraciones de 400 fibras/ml. El aumento de mortalidad por toxicidad pulmonar se observó a la mayor de las concentraciones, lo que sugiere que se había sobrepasado la Dosis Máxima Tolerada. Existe una considerable polémica acerca del potencial biológico de esas lesiones y su trascendencia para la especie humana. Ese estudio, por haber durado sólo 24 meses, tal vez no haya revelado todo el potencial carcinógeno de las fibrillas.

La instilación intratraqueal de una sola dosis de papel Nomex troceado (2,5 mg) que contenía fibras de diámetros comprendidos entre 2 y 30 μm provocó una respuesta inflamatoria inespecífica; dos años después de la exposición tuvo lugar una respuesta granulomatosa. La instilación intratraqueal de una sola dosis de 25 mg de Kevlar provocó una respuesta inflamatoria inespecífica que remitió al cabo de una semana aproximadamente; más tarde se observó una respuesta granulomatosa y un grado mínimo de fibrosis.

En tres estudios en que se inyectaron intraperitonealmente fibras de Kevlar (hasta 25 mg/kg) se observó una respuesta granulomatosa, pero no así un aumento significativo de la incidencia de neoplasias. Los autores de esas investigaciones indicaron que si no se había observado una respuesta neoplásica era posiblemente porque se había producido una aglomeración de las fibrillas de Kevlar en la cavidad peritoneal.

No existen estudios en que se haya examinado adecuadamente el potencial fibrógeno o carcinógeno de las fibras de poliolefina. En un experimento realizado con ratas sometidas durante 90 días a inhalación de fibras de polipropileno (hasta 50 fibras/ml) respirables (46% < 1 μm) se observaron cambios dependientes de la dosis y duración de la exposición, consistentes en un aumento de la celularidad y bronquiolitis. No se dispone de datos pertinentes sobre los efectos de la instilación intratraqueal. En los estudios efectuados en ratas mediante inyección intraperitoneal de fibras o polvo de polipropileno no se observó ningún aumento significativo de la incidencia de tumores peritoneales.

Los datos de que se dispone para evaluar la toxicidad y genotoxicidad *in vitro* de las fibras orgánicas sintéticas no son adecuadas. En el caso de las aramidas, los estudios han demostrado que las fibrillas cortas y finas de para-aramida tienen propiedades citotóxicas. Respecto a las fibras de poliolefina, hay algunos indicios de que las fibras de polipropileno son citotóxicas. Las pruebas de mutagenicidad realizadas con extractos de gránulos de polietileno arrojaron resultados negativos.

6. Efectos en el ser humano

En un estudio transversal realizado en 88 de los 110 trabajadores de una fábrica de producción de fibra de carbono de filamento continuo a partir de poliacrilonitrilo, la exploración radiográfica y espirométrica y los cuestionarios sobre síntomas respiratorios no pusieron de manifiesto ningún efecto respiratorio nocivo. En otros estudios no tan bien documentados se notificó el hallazgo de efectos adversos en trabajadores dedicados a la producción de fibras tanto de carbono como de poliamida; los datos publicados sobre estas investigaciones, sin embargo, eran insuficientes para determinar la validez de las correlaciones señaladas.

7. Resumen de la evaluación

La información sobre los niveles de exposición a la mayoría de las fibras orgánicas sintéticas es limitada. Los datos disponibles indican en general que los niveles de exposición en el entorno ocupacional son bajos. Cabe la posibilidad de que en futuras aplicaciones y usos se alcancen exposiciones más elevadas. No se sabe prácticamente nada sobre lo que ocurre con esos productos en el medio ambiente ni sobre su distribución o las exposiciones de la población general.

A tenor de los escasos datos toxicológicos obtenidos con animales de laboratorio, cabe concluir que existe la posibilidad de que la exposición ocupacional a esas fibras orgánicas sintéticas por inhalación tenga efectos adversos en la salud. El riesgo potencial para la salud asociado a la exposición a las fibras orgánicas sintéticas presentes en el entorno general se desconoce por el momento, pero probablemente sea muy bajo.